Crop Diversification in Jammu and Kashmir

THE AUTHORS

Dr. Banarsi Lal is working as an Incharge, Sr. Scientist-cum-Head at Krishi Vigyan Kendra (KVK), Reasi of Sher-e-Kashmir University of Agricultural Sciences and Technology-Jammu, India. He did his Masters in Extension Education from RAU, Bikaner and Doctorate from Banaras Hindu University (BHU), Varanasi. He has done extensive work on "Development of communication model on farm education programmes of television in Kathua district of Jammu and Kashmir" from 2002-2007 during his Ph.D. period. Dr. Banarsi has won gold medal in M.Sc. and topped in qualifying exam for Ph.D. programme conducted by Institute of Agricultural Sciences, BHU, Varanasi. He has completed his M.Sc. through ICAR exam. From the last thirteen years he has been working in KVK, Reasi with full dedication for the farmers' community and has wide experience in extension. He adopted an organic village Talwara and did extensive work in organic farming. He is upgrading the knowledge of farmers on organic farming so as to produce the safe and healthy food. So far he has published more than 25 high value research papers, 5 books, more than 500 popular articles, 50 technical bulletins and seventeen success stories/case study/ training manual. He is working as PI and CO-PI of more than 5 research/extension projects undergoing in the district. He is recipient of more than 40 zonal, national and international awards. Presently he is an active member of many Professional Societies.

Dr. Pawan Kumar Sharma is working as Scientist, Agricultural Economics in the Directorate of Extension at Sher-e-Kashmir University of Agricultural Sciences and Technology of Jammu. After completing graduation from Rajasthan and Masters from Agra, he acquired his PhD. degree in Agricultural economics from SKUAST-Jammu. Besides, having Post Graduate Diploma in Agribusiness Management, he has completed certificate courses in 'Agreement on Agriculture' and 'Sanitary and Phytosanitary measures' from World Trade Organization. He has more than 13 years of experience in agricultural research, teaching and extension. He has to his credit more than six books, thirty research papers and twenty five book chapters. He remains associated with more than five national and international research projects in the capacity of Principal Investigator and Co-PI.

Dr. Amit Singh Charak at present Incharge Sr. Scientist & Head KVK-Doda of SKUAST-Jammu, acquired B.Sc. (Ag.) Hons. from M.D.S. University, Ajmer, Rajasthan. He did his Master's as well as PhD degree in Agronomy from MPUAT-Udaipur, Rajasthan. He has published more than 40 reputed research papers and more than 100 popular articles in national & International journals and Magazines of well repute. He has received fellowship of reputed society and number of reputed National as well as International level awards/recognitions. At present, he is member of about 8 National/ International level professional societies.

Crop Diversification in Jammu and Kashmir

– Authors –

Banarsi Lal

Pawan Kumar Sharma

Amit Singh Charak

2022

Daya Publishing House®

A Division of

Astral International Pvt. Ltd.

New Delhi – 110 002

ISBN: 9789354614330

Published by : **Daya Publishing House®**
A Division of
Astral International Pvt. Ltd.
– ISO 9001:2015 Certified Company –
4736/23, Ansari Road, Darya Ganj,
New Delhi-110 002
Ph. 011-43549197, 23278134
e-mail: info@astralint.com
Website: www.astralint.com

Preface

Union Territory of J&K is endowed with ample natural resources including soil, water, climatic condition, diversity, topography, rich natural flora *etc.* which are conducive for the cultivation of a wide range of crops. There is dire need to increase the income and employment in agriculture and allied sectors for the farmers in J&K. There is possibility to increase the production and productivity of the crops and to enable the farmers to diversify their crop production in J&K. The crops production can be increased by adopting the modern technologies in agriculture and establishing the infrastructure for the farm production in J&K. There is need to promote the diversification in agriculture in J&K and move towards high value crops like vegetables, aromatic and medicinal plants, saffron, mushroom, bee-keeping, silkworms, pulses *etc.* The book entitled as "Crop Diversification in J&K" provides a comprehensive reading material for researchers, students, planners and extension field functionaries. It will also assist them in providing knowledge on diversified farming to the farmers. This book can also help the research scholars to further enrich the subject of agriculture. The book provides wider coverage of different aspects such as potential of organic farming, horticulture, floriculture, sericulture, commercial poultry farming, bamboo cultivation, bee-keeping, women empowerment and Self Help Groups (SHGs), mushroom cultivation *etc.* in J&K. There has been a long felt need for a book exclusively on "Crop Diversification in J&K. Endeavour has been made by the authors to collect information on various aspects of diversified agriculture in J&K. The thrust of the book is on crops diversification in J&K. The authors possess a vast research and field experience in the field.

A deep sense of respect and gratitude to Prof. J.P. Sharma, Hon'ble Vice Chancellor of SKUAST-J, Dr. S.K. Gupta, Director Extension, SKUAST-J,

Dr. Vikas Tandon, Dr. Suja Nabi Qureshi and Dr. Ajay Gupta for their technical guidance and help in bringing out this book. They always encouraged and gave us their moral support.

Valuable suggestions from the readers are always welcome for further improvement of the book.

Dr. Banarsi Lal

Dr. Pawan Kumar Sharma

Dr. Amit Singh Charak

Contents

Chapter 1

Potential of Diversified Agriculture in J&K

Agriculture is the backbone of Union Territory of Jammu and Kashmir and this sector provides employment directly or indirectly to around 70 per cent of this UT population. Agriculture contributes about 65 per cent of J&K revenue which signifies the over-dependence of the J&K on agriculture. Jammu and Kashmir is a mountainous UT in which about 30 per cent of the area is under cultivation. This sector contributes about 65 per cent of J&K revenue which signifies the dependency of J&K on agriculture. Jammu and Kashmir is located between 32.17 and 37.06 North latitude and 73.2 and 80.36 East longitude in the Himalayan region. The agro-diversity of the state varies from sub-tropical in Jammu region to temperate in Kashmir region. and cold arid in Ladakh region. There is dire need to increase the income and employment in agriculture and allied sectors for the farmers in J&K. This is possible to increase the production and productivity of the crops and to enable the farmers to diversify their crop production. By adopting the modern technologies in agriculture and establishing the infrastructure for the farm production. There is need to promote the diversification in agriculture and move towards high value crops like vegetables, aromatic and medicinal plants, saffron, mushroom, bee-keeping, silkworms, pulses *etc.* J&K is endowed with ample natural resources including soil, water, climatic condition, diversity, topography, rich natural flora *etc.* which are conducive for the cultivation of a wide range of crops. It is matter of grave concern that the interest of younger generation in J&K is dwindling towards agriculture. J&K 58 per cent area under agriculture is rain fed and remaining 42 per cent is irrigated. Maize is first rank crop in J&K occupying one-third of the total cropped area, rice at number second in 28.58 per cent area of gross cropped area and wheat at number three which covers around 26 per cent of the total cropped area.

In 2010-11 the food grain production of India for the year was 24156 thousand tonnes while it stands only 137.15 thousand tonnes in J&K reflecting a very low share of about 0.6 per cent. This trend is almost stagnated at around 0.6 to 0.7 per cent since a decade or so. It is pertinent to mention here that the food-grain production in J&K is not keeping pace with demand and there is a wide gap of about 5 lakh metric tonnes. J&K is mostly relying on imports mainly from Punjab and Haryana accounts for Rs 1600 crores. One of the strong reasons for low yield is lack of adoption of modern agricultural technologies and most of the region of J&K is rain fed. The net irrigated area in 1981-82 was 3.04 lakh hectares; it went up to 3.71 lakh hectares in 2008-2009 reflecting a very marginal increase even after three decades.

It is pertinent to mention here that the J&K could not qualify for National Food Security Mission (NFSM) - a flagship programme of Government of India. There is dire need of strenuous efforts to make agriculture more profitable and more income and employment can be generated among the farmers.

The horticulture sector in J&K accounts for 13 per cent of the net sown area but contributes 45 per cent of the state's agricultural GDP. Around six lakh families comprising of 30 lakh people are directly or indirectly involved in horticulture trade. Horticulture sector helps in nutritional security, land productivity, employment, exports and farm incomes by offering a wide range of choice to the farmers for crop diversification. The varied agro-climatic zones in J&K make it ideal to grow diverse high value horticultural crops. The forward and backward linkages offer tremendous potential for employment generation and rural upliftment in J&K. Horticultural sector contributes immensely to strengthen the financial condition of Jammu and Kashmir. This is a core sector of Jammu and Kashmir agriculture and about 4.5 lakh families are engaged in it. Around 20 per cent area of J&K is under horticultural crops. Horticultural sector contributes around 5000 crores to the annual income of Union Territory of Jammu and Kashmir. Among the horticultural crops in J&K apple occupies the predominant position constituting around 45 per cent of the total area under fruit crops. Horticultural activities major thrust is given in hilly areas. The fruit crops grown in J&K are almonds, walnuts, pears, cherries and apricots in temperate areas and mango, citrus, litchi, ber, papaya, guava *etc.* in subtropical areas. Saffron is the costliest spice of the world. Saffron cultivation in Jammu and Kashmir is unique in the world. The annual production of Saffron in Jammu and Kashmir ranges from 10 to 20 tonnes. J&K government has made strenuous efforts to increase horticultural crops production. Incentives are given to the horticultural crops growers by the departments of J&K to develop orchards in Kandi and hilly areas. Subsidy to the extent of 50 per cent of fruit plants and inputs used by the farmer are provided. It has been observed that around 40 lakh tones of fruits are exported annually from J&K. This sector contributes significantly to the economic and ecological development, employment generation, export and nutritional requirement of the people of state. Since Jammu and Kashmir is endowed with different agro-climatic conditions and topography, the scope for strengthening various commercial horticultural activities is enormous.

Agricultural policies involve the government, different organizations and the farmers. The government aims in maximization of agricultural outputs on sustainable basis. The agricultural universities and research institutions give policy inputs and the government departments prepare the action plans on the basis of programmes and make efforts to achieve the objectives of the government. The farmers make endeavors to maximize their farm income and employment. The planning policies which are top down in approach decide on macro-economic basis,

assuming that the farmers would adopt them or could be induced to adopt. The general policy of government for agricultural development can be summarized in terms of (a) supply of inputs like High Yielding Varieties (HYS) seeds, fertilizers and insecticides-pesticides at reasonable rates;(b) supply of water, credits and electricity at subsidized rates; and (c) fixing of minimum support price for important food grains and other crops along with procurement system for wheat and rice. The farmers have responded to these policies by allocating higher acreage under wheat and paddy crops which have least risk of yield as well as price. In the nineties after reaching plateau in wheat and rice yields, some progressive farmers started to increase the production of floriculture crops, vegetables and fruit crops, citrus, strawberry and even in mushroom cultivation. But the farmers did not get the proper policy for these commodities' which caused them heavy losses and they again were forced to cultivate the wheat and rice crops. The policy of diversification has been loaded with objectives like checking degradation of soil health, sustaining the land productivity, checking decrease in level of water table and waterlogging, controlling the use of insecticides and pesticides, decreasing the water pollution, maintaining the ecological balance and increasing production of crops. The farmer thinks for the short term benefits. The continuous increase in area under wheat and rice in spite of echoing of the above mentioned considerations clearly proves the fallacy of this policy of diversification.

The support price with procurement arrangement has been revealed as most important instrumental variables in influencing the acreage allocation decisions of the farmers. This support is becoming unsustainable from the government. The support price with procurement arrangement for additional crops is almost ruled out in view of prevailing weak fiscal situations of the central and state Governments. The acreage planning for individual crops is widely used as a management tool by almost all the developed countries. Acreage planning requires realistic estimates of the acreage under individual crops at national level on the basis of domestic and export demand of the concerned crops. The level of production and acreage can be fixed for different crops at national level so as to fetch a market price reasonably above the cost of production.

Apex organizations in State like SKUAST-J, SKUAST-K and Central Institute of Temperate Horticulture- (CITH) -Srinagar have made commendable achievements through research and extension but still there is an immense scope for further improvement to increase production and productivity of different crops in J&K. Efforts need to be made for further exploration of high yielding varieties, production of quality planting material, canopy management, high density plantation of fruit crops, rejuvenation technology and water harvesting. Different state and centrally sponsored schemes/programmes can change the fate of agriculture scenario in the state. Emphasis should be given on crop husbandry as food security and diversification are indispensable. Technologies developed through Front Line demonstration (FLDs) and On-Farm-Trials (OFTs) should be disseminated to the farmers through main extension system of the state. There is need to change the outlook of the farming community form traditional agriculture to diversified agriculture.

Chapter 2

Agro Techniques for Hilly Areas Development

In technology development, technology choice has a critical impact on many aspects of hilly areas development, especially the way we choose it, the way we design it, the way we deliver it to the masses. Technology is carefully chosen to enable the local people to imbibe knowledge appropriate to their needs and environment. Technology should upgrade the skills and capabilities, reduce drudgery, minimize fatigue, capable of easy assimilation, generate added value to the existing methods of operation, generate employment, low in capital investment, low in cost of production, be capable of replication and adoption and should blend harmoniously with existing ecosystems leading to tangible improvements in the living conditions and development of the people of hilly areas. There is a need to introduce appropriate technologies in the hilly areas particularly in disadvantaged groups. There is a need of institutional linkages and active participation of voluntary organisations, Science and Technology based institutions, Research and Development institutions, financial agencies and most importantly people who are the primary stakeholders. Success of these technologies lies on participatory systems with a systematically approach for effective dissemination of technology. There should be proper mechanism in development and transfer of effective technologies in the hilly areas. The requirement of technology is assessed by the Science and Technology based institutions and NGOs. These organisations then look for ideal technology option ensuring the availability of local resources and skills. Technology is developed according to the in-house capability of the local people with the assistance of technical institutions. The technology is developed in such a way so that it is accepted by the people of hilly areas and which is best suited in their local conditions for a long-term sustainability. The people are trained for 2-3 years to handle the new technology and they are also trained in its maintenance and repair. The Science and Technology institutions and NGOs also helps in backward and forward linkages both for production as well as marketing for long-term sustainability.

Looking into constraints and priority needs of the people of hilly areas, various thrust areas need special attention ensuring involvement of local community through network of developmental field groups in technological empowerment. Agriculture in hilly areas is the most important activity as it is directly related with the villager's sustenance. Agriculture in hilly areas is mostly rainfed, poor crop productivity, poor cultivation practices *etc.* The interventions in hilly areas should be viewed as specific as follows:

- ☆ **Organic farming**: Organic farming has immense potential in hilly areas. The litter fall in hilly areas can be utilised for compost making. This can highlight unproductive hilly areas contour as organic farm. Organic products produced in hilly areas can help the farmers to fetch better prices.
- ☆ **Off season vegetable and seed production:** In hilly areas off season vegetables can be grown which can fetch higher prices. Seed production and low cost nurseries units made in hilly areas can generate the income.
- ☆ **Non-perishable cultivation:** The perishable products do not fetch better prices due to various market intricacies. The hilly areas agro-climate offers better potential for non-perishable products such as spices, pulses, oilseeds *etc.*
- ☆ **Horticulture:** Horticulture has lot of scope in hilly areas. Different varieties of fruit crops can be produced in hilly areas. But the fate of all such produce has been poor due to certain reasons such as: transportation facilities, storage facilities and processing facilities. Horticulture in hilly areas is largely ignored but it has immense potential. Hilly areas farmers need proper training for horticultural crops. There should be storage, processing and packaging facilities of fruit crops for commercial purpose.
- ☆ **Animal husbandry:** In hilly areas scarcity of fodder crops for animals is observed specially in off seasons. There is a need to produce high breeds of animals in hilly areas. Buffalo rearing, poultry farming and goat rearing have lot of scope in hilly areas. The special varieties of fodder crops should be introduced in hilly areas.
- ☆ **Cover Management:** Green cover in hilly areas is very important for the maintenance of environment. The green cover with appropriate bio-engineering plantation management is considered as need based.
- ☆ **Water power:** Water and energy are important resources for the hilly areas ecosystem. In hilly areas large numbers of dams are in process in but the traditional water mills, huller, carding devices should also be taken into consideration for the up gradation. Strenuous efforts are

needed to develop sustainable technologies for harnessing and conserving renewable energy like water lifting devices, solar lantern *etc.*

- ✰ **Disaster Management:** Capacity building and technological empowerment are needed to mitigate the disaster management challenges for the people of hilly areas.
- ✰ **Herbs:** Medicinal plants occurring in hilly areas have a great national and international importance. Growing awareness towards herbal therapy provides immense scope for developing herbal industries in hilly areas.
- ✰ **Health care and sanitation:** Health care of hilly areas people is an important area that needs to be taken care of. There is need of technology intervention in the area of rural sanitation.
- ✰ **Tourism:** Tourism forms the basis for the economy of many hilly places. There should be linkage between the people of hilly areas and hill tourism. Rural tourism should be promoted in hilly areas.

It is concluded that the bottom up approach for participatory technology development starting with the people and with the contribution of Science and Technology Institutions and close interface with voluntary organisations can provide sustainable models for technology development and transfer. Appropriate networking and linkages amongst people and institutions can change the lives of the people of hilly areas through technological empowerment and capacity building at the grassroots level.

Chapter 3

Horticulture for Rural Prosperity in J&K

Cultivators are said to be very important citizens on this earth because they feed the world. They are the most independent, vigorous and virtuous individuals. India is an agricultural country with 143 million hectares of land as net sown area, the highest percentage of land under cultivation in the world. The country accounts for 17 per cent of world's population and 2.4 per cent of world's area. Horticulture sector encompasses a wide range of crops such as fruit crops, vegetables crops, ornamental crops, spices, medicinal and aromatic crops *etc.* Horticulture is the fastest growing sector in India and contributes immensely in poverty eradication and nutritional security. This sector has immense scope in India to increase the income and employment for the population and helps in sustaining large number of industries. Horticultural crops play a unique role in the Indian economy by improving the income of the farmers. Cultivation of these crops is labour intensive and lot of employment is generated for the rural people. India is the second largest producer of fruits and vegetables in the world after China. India contributes 10 and 13.4 per cent of fruits and vegetables in the world. India is the largest producer of mango and banana in the world producing 65 and 11 per cent of mango and banana of the world. India is also the largest producer of cashew in the world. India is the largest producer of cauliflower in the world. India is second in onion production and third in cabbage production in the world. India's varied climate is suitable for the cultivation of more than 30 fruit trees and 40 vegetables. Horticulture covers only 6.1 per cent of gross cropped area but it contributes 18.8 per cent of gross value of agricultural output. Thus, horticulture is important in creating the employment, enhancing the profits and providing the nutritional and economic security to the farmers. It is estimated that 6.7 per cent growth rate in horticulture sector is must in order to achieve 4.0 per cent growth rate of agriculture. Government of India has strengthened the research and development activities in the horticulture sector. There is an increase in the demand of fruits and vegetables. So there is an urgent need to increase the production of fruits and vegetables. During the last decade has introduced certain technologies like greenhouse cultivation, micro-irrigation, precision farming, post-harvest management *etc.*

Horticultural sector contributes immensely to strengthen the financial condition of Jammu and Kashmir. This is a core sector of Jammu and Kashmir agriculture and about 4.5 lakh families are engaged in it. Around 20 per cent area of Union Territory of J&K is under horticultural crops. Horticultural sector contributes around 5000 crores to the annual income of the Union Territory of Jammu and Kashmir. Among the horticultural crops in J&K apple occupies the predominant position constituting around 45 per cent of the total area under fruit crops. Horticultural activities major thrust is given in hilly areas. Around 45 per cent of economic returns of agriculture come from horticultural sector that contributes around 8 per cent of this UT's economy. The fruit crops grown in J&K are almonds, walnuts, pears, cherries and apricots in temperate areas and mango, citrus, litchi, ber, papaya, guava *etc.* in subtropical areas. Saffron cultivation in Jammu and Kashmir is unique in the world. The annual production of Saffron in Jammu and Kashmir ranges from 10 to 20 tonnes. The J&K government has made strenuous efforts to increase horticultural crops production. Incentives are given to the horticultural crops growers by the departments of this UT to develop orchards in Kandi and hilly areas. Subsidy to the extent of 50 per cent of fruit plants and inputs used by the farmer are provided. It has been observed that around 40 lakh tones of fruits are exported annually from J&K. This sector contributes significantly to the economic and ecological development, employment generation, export and nutritional requirement of the people of J&K. Since Jammu and Kashmir is endowed with different agro-climatic conditions and topography, the scope for strengthening various commercial horticultural activities. In horticultural crops post -harvest losses range from 15-20 per cent. In J&K around 94 per cent of the land holders fall in the category of small and marginal farmers.

There is an urgent need to focus our research and development efforts on increasing the production, productivity and quality of fruits and vegetables, besides paying undivided attention to reduce the cost of production. Declining shares of fruits and vegetables in terms of area and production and exports in horticulture and plantation crops need to be arrested. We should utilize our research and development to increase the fruits and vegetables production. In order to increase the production, productivity and quality of produce and reduction in the cost of production, our research institutes/centers have identified certain key areas and have generated current knowledge in this area which needs to be transferred to the farmers in a systematic way so that farmers are enabled to increase the area under fruits and vegetables, adopt the scientific technology, use the optimum inputs and achieve the demand of fruits and vegetables. Production of fruits and vegetables can be increased by increasing the area and by applying the scientific

production technology. It is very difficult to increase the area under these crops but still there is possibility to increase the area by adopting the methods such as (a) By utilization of land through inter-cropping/mixed cropping in existing orchards, growing crops in the vacant areas and growing shade loving crops. (b) Development of appropriate varieties of crops and agronomic practices to suit reclaimed wastelands and dry land region in the country. (c)Adoption of cost effective poly house technology. (d) High density planting by reduction in planting by reduction in planting distance or by use of plant growth inhibitors and dwarfing rootstocks in crops like mango, citrus, banana, apple and some other fruits. (e) Use of protected cultivation under controlled conditions for growing fruits like strawberry and some vegetables. Productivity can be increased by bringing maximum possible area in a systematically drawn plan of five years under high yielding or rejuvenating the already existing orchards along with full package of scientific practices. This will help in improving the quality of produce and reduction in the cost. In the conventional methods of crop improvement both desirable and detrimental genes get transferred from parents. The modern biotechnology system helps to insert single or more than two desirable genes into the crop. Traits can be modified through the genetic engineering that helps to increase productivity and improve the nutritional quality. This technique helps production of uniform size, shape, size and colour of products. Micro-propagation is the most popular technique for production in the world and supply of identical plants. Propagation technique through tissue culture and molecular indexing for culture of diseases are immense use in making available healthy planting material. Micro propagation protocol has been developed in banana, grape, ginger, turmeric, strawberry and betel vine. Farm mechanization now-a-days is very much helpful in reducing the labour cost and conducting timely farm operations and efficiently. For mechanizing cultivation of some fruits and vegetables some implements have been designed for distribution to the horticultural crops growers. The implements such as mango harvesters, potato diggers, automatic potato planter, potato ridger, potato grader, soil crust breaker, fruit peeler, fertilizer application-cum-line marker *etc.* are very important for the horticultural crops growers. These equipments must be widely demonstrated to the fruit growers to establish their suitability, cost-effectiveness *etc.* and then makes them available.

Integrated pest and disease management aims at judicious use of plant protection methods *viz.* cultural, biological, chemical, physical/mechanical or regulatory control methods. Biological control assumes significant importance and calls for the development of natural enemies tolerant to multi-pesticidal groups. Integrated nutrient management refers to maintenance of soil fertility and plant nutrient supply to an optimum level for sustaining the desired crop productivity. It aims at achieving the highest level of productivity with efficiency at economically acceptable limits in environment friendly way. Use of organic manures and biological fertilizers helps in restoring soil health, reducing quantity and cost of chemical fertilizers. Awareness on biological nitrogen fixation, phosphate solubilizers and micorrhizal fungi must be created among the farmers. Micro-irrigation and fertigation techniques must be popularized among the farmers. Under this system, water is applied at a low rate for a longer period at frequent intervals. It has been successively tried in banana, mango, pomegranate, citrus, cucumber, okra, brinjal *etc.* Now emphasis must be given to apply the fertilizers through micro-irrigation. Micro-irrigation and fertigation helps to save the fertilizers up to 30 per cent and save water up to 40 percent. Moreover, this system helps to prevent the growth of weeds, saves energy and improves quality of produce. Greenhouse technology is a capital intensive, sophisticated and high–precision technology which facilitates intensive cultivation of vegetables throughout the year. Greenhouse technology can modify the environment for optimum plant growth and production of uniform and high quality produce. Fan and pad cooled greenhouses are suited for hot and dry climate and heating system greenhouses are suited in cold climate. In greenhouse technology wind and thermal buoyancy obviates the need for electric power or generators for maintenance of temperature and relative humidity. This technology should be demonstrated to the Indian farmers. This technology is more suited for the cultivation of vegetables in northern hilly regions during summer season. Organic farming contributes in improving the quality of fruits and vegetables which have potential export market. Major components of organic farming are addition of organic manures, adoption of crop rotations and enhancement of soil fertility through biological fixation of nitrogen. Crop residues are utilized to make compost along with farm-yard manure and biogas slurry. Incorporation of green manure or green leaf manure helps to convert the crop residues into the organic matter in the soil. Insects-pests can be controlled by the combination of crop rotations, inter-crops, mechanical cultivation and use of biological pest control. Proper post-harvest management reduces the post-harvest losses by proper handling and packaging and adoption of suitable low cost storage infrastructures. Use of pre-cooling units, controlled atmosphere and refrigerated containers is recommended to increase the shelf life of horticultural crops. Indian farmers need to improve the quality of crops to make them qualitatively competitive in world market. There is need to establish the quality control laboratories to ensure the exporters that commodities being exported get international standards.

In order to boost the horticulture crops production state governments, need to prepare a vision and mission on horticultural crops, define states role clearly and mobilize the financial resources to invest in this sector. Agricultural universities should take lead role in providing the proven technical guidelines to the horticultural crops growers and demonstrate effectiveness of technologies on growers' fields for adoption. Adequate and timely credit aid by financial institutions/banks is called through proper estimation of credit at growers'/ village level, assisting self-help groups, providing Kisan Credit Cards, providing post-harvest facilities and linking the growers with the Agricultural Universities for demonstration of proven technology.

Chapter 4

Climate Variability and Fruit Crops Production in J&K

Union Territory Jammu and Kashmir is endowed with different agro-climatic conditions and topography. There is enormous scope for strengthening various commercial horticultural activities to enhance the income of the farmers. Horticultural sector contributes immensely to strengthen the financial condition of J&K. This is a core sector of Jammu and Kashmir agriculture and about 4.5 lakh families are engaged in it. Around 20 per cent area of the UT is under horticultural crops. Horticultural sector contributes around 5000 crores to the annual income of Jammu and Kashmir. Among the horticultural crops in the UT apple occupies the predominant position constituting around 45 per cent of the total area under fruit crops. Horticultural activities major thrust is given in hilly areas. The fruit crops grown in this UT are almonds, walnuts, pears, cherries and apricots in temperate areas and mango, citrus, litchi, ber, papaya, guava *etc.* in subtropical areas. Saffron cultivation in Jammu and Kashmir is unique in the world. The annual production of Saffron in J&K ranges from 10 to 20 tonnes. The J&K government has made strenuous efforts to increase horticultural crops production. SKUAST-J and SKUAST-K has made great efforts to introduce new interventions in horticultural sector in J&K. Incentives are given to the horticultural crops growers by the departments of the UT to develop orchards in Kandi and hilly areas. Subsidy to the extent of 50 per cent of fruit plants and inputs used by the farmer are provided. It has been observed that around 40 lakh tones of fruits are exported annually from J&K.

Horticultural sector contributes significantly to the economic and ecological development, employment generation, export and nutritional requirement of the people of UT. Horticultural crops are adapted to a wide range of climatic conditions, produce higher biomass per unit area as compared to field crops, are more remunerative for replacing subsistence farming and thus may aid in alleviating poverty in difficult agro- ecosystems as rain fed, dry land and hilly areas of J&K. There is also great potential to improve wastelands for productive use in the UT. These crops require lesser irrigation and provide higher employment opportunities round the year. Increase in demand for horticultural produce due to great health awareness, rising income, export demand and increasing population poses the challenge for further increasing the production and productivity of horticultural crops. The issue of climate change and climate variations has resulted in more uncertainties and risks in this already high capital intensive system. This has resulted in further aggravating the constraints on horticulture production system. The major challenges before us is to have sustainability, higher levels of production, competitiveness to stay in market, regular production, land, water and more importantly threat of climate change. In order to meet the above challenges, we need innovations for improving horticultural crops production in new era.

The earth's climate makes small adjustments so as to maintain a firm balance between the energy reaching earth surface and energy reflected out of it. Even a small rise in temperature could mean substantial changes in cloud cover and wind pattern. Some changes may enhance warming or some may counteract warming effects. Increase in small particles of matter or liquid produced naturally or by man-made activities known as aerosols results in cooling effect. Increase in water vapour due to increase in evaporation due to high temperatures can further add to rise in warming effect. The significant change may impact horticultural crops production. Climate change results in erratic rainfall pattern, heat waves, more warm spells and crop failures. However, climate change may have some beneficial effect to a limited extent. The vulnerability of horticulture depends on both expected regional climate change and that sector's ability to adapt to the change. The projected or anticipated increase in temperature, variability in precipitation pattern and increase in frequency of extreme weather events such as heat, cold waves, frost days, drought and floods would severely affect the production of horticultural crops. This stress at different crop development phases in varying intensities would ultimately determine productivity and quality.

The Indian climate is dominated by monsoon, responsible for most of the regions precipitation, poses excess and limited water stress conditions. Fruit crops being sensitive and succulents are generally more sensitive to environmental extremes. High or low temperatures and water stress are the main cause of low yields. Under climate change scenarios the impact of these stresses would be compounded. The horticultural crops having C3 photosynthetic metabolism have shown beneficial effects due to elevated carbon dioxide. The extreme hot and cold wave conditions have been reported to cause considerable damage to many fruit crops. In perennial, crops like mango and guava, temperature is reported to have influence on flowering. The percentage of hermaphrodite flowers is more in late emerging panicles, which coincided with higher temperatures. Though grape originated in temperate regions, modifications in production system has enabled

it to adapt to subtropical and tropical climates. Under climate change conditions there would be changes in availability of growing degree days leading to hastening of the phonological processes. Most of the subtropical fruits are grown in rain fed conditions and are vulnerable to climatic variability and drought conditions caused due to shifts in rainfall pattern and temperature fluctuations.

The rise in temperature will influence the survival and distribution of pest populations. The rise in temperature increases nutrient mineralization in soils which will decrease fertilizer use efficiency. Increase in temperatures at fruit maturity stage results in rise in events as fruit cracking and burning of litchi and premature ripening in mango. Low fruit setting is observed due to sudden rise in day temperatures (>35°C) during peak flowering in mango. Low temperatures

(4-11°C) in January, accompanied with high humidity (>80 per cent) and cloudy weather delays panicle emergence. During peak bloom period, high temperature (35°C) accompanied by low humidity (45 per cent) and long sunshine hours result in excessive transpiration and dehydration of panicles. Untimely winter rains promote vegetative flushes in citrus instead of flowering flushes. Dry spell during flower emergence and fruit setting affects flower initiation and aggravates pest incidences. In wine grapes, anthocyanin development is influenced by difference between day and night temperatures with high variations (15-20°C) promoting colour development.

Many slow growing fruit crops require heavy investment on establishment of orchards. Changing the existing varieties would be difficult and very costly under the impact of climate change. In Kullu district of Himachal Pradesh, farmers have shifted from apple cultivation to either pomegranate cultivation or to vegetable growing. While, in Shimla district at relatively higher altitudes orchards have been replaced from high chilling requiring apple cultivars (Royal Delicious) to low chilling requiring cultivars and other fruit crops as pear, plum and kiwi *etc.* In mid hills the trend is altogether shifting from apple cultivation to vegetable cultivation due to erratic weather conditions and reducing chilling hours. It is corroborated by declining trend in snowfall and apple productivity in Himachal Pradesh and Jammu and Kashmir. Thus in replanting orchards and in new plantations during these years, selection of low chilling types may be advisable. This is an example of impending impact of climate change. Hence, under such circumstances, we would have to identify varieties and regions suitable for production of quality fruits. With global warming production areas for specific crops and/or time of planting could be changed. For many horticultural crops areas have been defined for fruit crops and creating new infrastructures in new areas would not be an easy task. Location of important production area is often defined as much by available land, markets and infrastructure as by climatic conditions. Thus, climate change and CO_2 are likely to alter important interactions between horticultural plants and pollinators, insect-pests, diseases and weeds *etc.* Keeping in view the nature of crop, its sensitivity levels and the agro-ecological regions in J&K, the crop based adaptation strategies need to be developed, integrating all available options to sustain productivity. To prepare ourselves for climate change and for formulating a sound action plan we must identify the gaps in information, prioritize researchable areas, and formulate concrete policies.

Chapter 5

Commercial Beekeeping in J&K

Honey is a great source of natural energy. It has an incredible medicinal properties. Bee-keeping can be a profitable enterprise, if it is done scientifically. Besides honey, pollen, beewax, royal jelly and pollination are other revenue generating sources of bee-keeping. Bee-keeping increases the pollination process and increases the crops production. It is estimated that about 70 metric tonnes of honey is produced in J&K. Union Territory of J&K has tremendous potential in bee-keeping because of its conducive climate for bee-keeping. It is endowed with diverse forest resources. A good area under forests and negligible amount of agro-chemicals use in agriculture in J&K increases the potential of organic honey production. Horticulture sector in J&K generates around 1500 crores of revenue and thus sustains its economy and bee-keeping boosts the horticultural sector by intensifying the pollination process. Horticultural sector in J&K provides employment to around 25 lakh people and supports to around 6 lakh families. Bee-keeping helps to generate more income and employment as it increases the crops production by 15-20 per cent. It has been observed that almond production can be enhanced by 33-47 per cent by providing adequate number of pollinators. Bee-keeping has positive impact on both horticulture and agriculture. It improves the overall production and quality of crops. It has been observed that in some areas fruit crops production has declined due to lack of pollinators. More emphasis should be given on the pollination especially in case of fruits crops so that the quality and quantity of fruits crops production can be enhanced. Bee-keeping can play a significant role in increasing the crops production in these areas. By bee-keeping crops production can be enhanced even in less irrigated areas. It has been observed that a unit of 10 colonies multiplies to about 15-20 colonies in a year. Beewax and honey can be a very good enterprise to export and generate more money for the farmers. This venture has an immense potential in J&K and can bring prosperity among the farmers if explored commercially. J&K Government is promoting the bee-keeping and various organisations are coming forward to support the farmers for bee-keeping. Krishi Vigyan Kendras (KVKs) of SKUAST-J and SKUAST-K and Department of Agriculture are imparting technical skills among the farmers. Beehives are also provided to the farmers on subsidy basis so that bee-keeping can be promoted. Not a big infrastructure is required for establishing bee-keeping unit. This enterprise does not require big capital amount for start-up rather it can be initiated with little capital amount. It can be a remunerative profession for the rural youths. Beekeeping can be great source of livelihood for the farmers of J&K.

J&K is considered to be one of the most important and suitable bee-keeping places in India. Bee-keeping has been an age-old tradition in J&K and is practiced from generation to generation. For commercial bee-keeping, scientific techniques need to be applied. Special capacity building programmes need to be organised for the beekeepers. Unemployed youths of J&K can come forward for bee-keeping. Agricultural universities need more research on this venture so that it can be adopted at a large scale. Pollination centres need to be established and beehives can be provided to the unemployed youths on rent basis. Bee-keeping can prove as a promising prospect in J&K. Bee-keeping can revolutionize horticultural sector in J&K. Unemployment is a serious problem in J&K and it seems that it is aggravating. Government jobs are very less and private sector is almost negligible in J&K. More number of educated youths are becoming unemployment in J&K. Unemployment problem in J&K can be mitigated through bee-keeping up to some extent. Bee-keeping can be a vibrant source of employment in J&K. This sector can play a significant role in doubling the farmers' income by 2022.The value addition in honey can bring a boost in J&K and can further generate more employment in J&K. Testing, bottling, packaging, storing, branding and marketing of honey can bring prosperity in J&K. Fabrication of gloves, bee veils, queen excluders, honey extractors, smokers *etc.* can also open new avenues of employment in J&K. Sometimes farmers are jolted by the parasitic mite (Varroa destructor) which kills bee colonies resulting into losses to the farmers. Timely management of this mite is essential to protect the hives.

Agricultural crops are seasonal and provide bee forage for limited periods only. Bee colonies cannot be sustained throughout the year in any cultivated area, unless there has been an integrated intensive agriculture, agro-forestry and social forestry systems. During the forage scarcity periods between two crop seasons, bee colonies need to be moved to another area. J&K possesses a variety of crops and tree species and thus present good opportunities for migratory bee-keeping. In J&K, colonies from Jammu region can be migrated to other places within the UT rich in flora during the off seasons. In J&K, the colonies are kept in different locations in the plains where mustard and toria are cultivated. During the severe cold in January, the colonies perform well even when they get very short foraging time. In the months of February and March colonies get nectar from mustard and eucalyptus. Honey yield can be increased in September from *Citrus* spp. In J&K,

mustards and eucalyptus flower later, *i.e.*, in February-March. Colonies can therefore be taken to other areas of UT. Colonies can then be migrated to litchi growing areas in March-April for the main flow from litchi. J&K is suitable for the stationary and migratory bee-keeping because of its favourable climatic conditions and diverse natural sources. Bee-keeping with *Apis mellifera* is popular in J&K. Knowledge of floral resources and appropriate migration schedules in different bee-keeping regions can increase the quality and quantity of honey. Various migration schedules for different phytogeographic regions in J&K are suggested by the scientists. Mustard, gram, eucalyptus, shisham, berseem, toria, maize, citrus, guava, cucurbits *etc.*, are grown in J&K and these crops are preferred by the bee colonies. Migratory bee-keeping in J&K can help to maximize honey yields and farmers can fetch more money. The natural flora of J&K forests is different from many other parts of the country. J&K has vast geographical area, varying climate from temperate to tropical, torrid to arctic and from total aridity to a maximum of humidity. Forests cover about 19.95 per cent of the area in J&K. Large quantities of deodar, fir, toon, teak and other trees are grown for timber. Several of these species provide nectar to honey bees. Oilseed crops like toria, mustard *etc.* are useful to honey bees. Among other commonly cultivated plant species useful to honey are mango, lemon, apple, orange, plum, litchi, pear, peach, cherry, jamun *etc.* Road-side plantations can also contribute to honey production in J&K. Public and private organisations support is needed for the farmers. After white revolution and green revolution extra efforts are needed to make sweet revolution successful. Sweet revolution can transform the lives of farmers of J&K.

Chapter 6

Poultry Farming as an Enterprise in J&K

Poultry farming is not new in J&K and it has been practiced for the farmers since the time immemorial. Most of the backyard poultry production in the state comprises the indigenous birds. The indigenous poultry birds produce only 60-80 eggs per year and meat production is also very less. Commercial poultry production can be easily boosted in the state with the introduction of improved varieties of poultry which can lay 160-200 eggs per year and also produce more meat. In J&K poultry farming is emerging an important enterprise and is assisting to create income and employment for the rural youths of the state. There is immense potential of poultry farming in the state. There are numerous examples of successful poultry entrepreneurs in the state. Government is offering the various incentives to motivate the aspiring poultry farmers of the state. Krishi Vigyan Kendras (KVKs) of J&K are imparting training programmes on advance poultry farming to the farmers and are creating interest on poultry farming in the state. KVK, Reasi has introduced black meat poultry variety namely Kadaknath in the district which is very low in cholesterol (0.73 per cent -1.05 per cent), rich in proteins (25 per cent), vitamins and helpful for the asthmatic and nephritis patients. The meat and eggs of this poultry variety are costly and can help to raise the socio-economic status of the farmers. KVK has also introduced high egg laying birds such as Vanraja, Chabro and Chabrown in Udhampur and Reasi districts. The Front Line Demonstrations (FLDs) laid by KVK, Reasi on Kadaknath, Vanraja and Chabrown have significant impact on the income of the farmers. Now many farmers of the area have become the poultry entrepreneurs. KVK, Reasi imparts regularly the short training courses to the interested farmers of the region.

Poultry farming has immense potential in the Union Territory of J&K. Despite the strenuous efforts poultry farming is not picking up pace in J&K. In J&K there is 69 per cent shortfall in the poultry meat and 97 per cent shortfall in the poultry egg production. The gap between demand and supply of poultry meat and eggs is filled by purchasing the poultry birds and poultry eggs from the other states of the country thus draining the J&K exchequer. According to the recent report, in J&K state every year around Rs.2000 crores of meat is imported from the outside the state. A little progress has been made on the broiler production in J&K but a poor progress has been made in commercial egg laying industry in J&K. Some outstanding egg laying poultry strains such as Vanraja, Chabro and Chabrown have been introduced in J&K which have raised hope for mitigating the eggs need in J&K. The cost of quality proteins is increasing rapidly and due to poor purchasing power of the farmers their families' members suffer malnutrition especially of animal protein. The poultry sector can ensure the proper nourishment of the farmers' families besides income and employment. Poultry birds can increase the income of the farmers as poultry farming requires low inputs, have short gestation period and give better and quick returns.

In the domestic market, the consumption of poultry meat has been low due to many reasons. The main reason is the low purchasing power of people. Only 25 per cent of the population living in urban areas consumes about 75-80 per cent of eggs and poultry meat. The per capita consumption of egg is 100 and poultry meat is 1.2 Kg per person per annum in urban areas. But in rural areas, it is only 15 eggs and 0.15 Kg poultry meat. Egg industry showed an excellent growth in the country due to (a) Egg layer industry large units with millions of birds in one house are coming up (b) 70 per cent of the egg laying birds are in the states of Andhra Pradesh, Tamil Nadu, Maharashtra and Karnataka. (c) fresh eggs are available at more reasonable costs.

Broiler industry grew in the country because (a) the commercial broiler chicks' ready in 60 days with tender meat. (b) Movement of parent stock, hatching eggs, day-old chicks initiated the broiler growing everywhere. (c)The broiler growing period has gradually come down to less than 40 days from 60 days due to advance technologies in the poultry industry. (d)Broiler rearing is giving excellent results. 2 kg broilers are produced on 36 days with 1.5 kg feed per kg chicken with less than 3 per cent mortality are being achieved on low- cost open houses reared by the small and marginal farmers. Around 60 per cent rural population in India has backyard poultry production. Poultry has been there in the backyards of most of the houses since ages, forming a part of nutrition and family income. Government has started encouraging the backyard poultry for the rural development. Improved varieties of low technology input birds are bred for this purpose, which are genetically more efficient in production in comparison to the local poultry. The birds grow faster than native birds reaching 1.5 kg weight in 45 days but can withstand variable feed qualities and also supplement themselves by scavenging. The low technology input birds are mostly of dual purpose. Mother units are being encouraged to take care

of the brooding and vaccination in the first four weeks before they are distributed among the farmers to ensure the disease security. Cluster farming in rural areas is possible with these chicken for enhancing the meat production on the lines of broiler farming in which of group of 8 to 10 farmers or educated youth can form a cluster and grow the birds in groups of 500 to 1,000 birds in low cost houses as "organic chicken" by making their own low-cost organic feeds without chemicals and antibiotics. The objective of backyard poultry is to encourage poultry farming activity and to provide employment opportunities in backward areas.

There are numerous opportunities in poultry industry in India due to (a) Government policies relating to investments in poultry and related industries, taxation, import duties, excise duties are favourable. (b) Increase in income and employment generation for the rural poor and marginal farmers. (c) Consumer awareness and acceptability of eggs and chicken as good source. The Indian Council of Agricultural Research (ICAR) is the nodal organization for agricultural research in the country that is playing vital role in this sector. It undertakes research in its own institutes Central Avian Research Institute, Izatnagar and Project Directorate on Poultry Hyderabad and also in state agricultural universities through All India Coordinated Research Projects.

There are certain challenges in the poultry industry such as: (a) Poor infrastructure for export is hindering the export of poultry products. (b) Competition from international players on opening up duty-free imports, lifting of trade barriers. (c) Increasing propaganda and demonstrations by organizations on promoting vegetarianism and animal rights. (d) Occurrence of Salmonella and other diseases in poultry meat. (e) Many countries are dumping their poultry products *i.e.* exporting eggs at prices lower than production cost. (f) Many countries are protecting their poultry industry from foreign competition by protective measures like restricting imports, keeping egg prices at lower level *etc.* (g) Some countries provide subsidies, export incentives to exporters and keep their price low. As the poultry industry is among the fastest growing in the world, it needs greater integration, better cost-effectiveness and improvement in the distribution. More retail outlets, mass gathering and creating awareness to home to home about the nutrient values of chicken and eggs is needed. Poultry industry needs good branding system in order to increase the consumption of chicken. The poultry companies have to encourage direct procurement of maize from the farmers by using contract farming. Also, the collection of reliable, updated statistics is necessary for immediate and long term planning and thereby helping in preventing shortages in poultry farming. There is need to develop efficient, independent, authority for disease monitoring, biological quality control and biosecurity measures in the poultry industry.

Chapter 7

Potential of Sericulture in J&K

Rearing of silkworms for the production of raw silk is known as sericulture. Sericulture involves the rearing of silkworm, collection of cocoons, reeling of raw silk from the cocoons and availability of mulberry trees from which fresh leaves can be obtained to feed the silkworms. J&K is one of the leading silk producing territories in India. In J&K there are around 2,800 villages where the sericulture is considered as an important activity for the farmers. In J&K there are around 7 lakh mulberry trees out of which 53 per cent are available in Jammu region and 47 per cent in Kashmir region. Approximately 6,680 quintals of raw silk worth about Rs.50 crores is produced annually in J&K. J&K is the only territory in the country which is at the same altitude in which leading biovoltine sericulture countries of the world lie. J&K is bestowed with the conducive atmosphere to produce the biovoltine silk. Sericulture is the traditional cottage industry in J&K. J&K produces high quality biovoltine silk comparable to the international standard. Sericulture can help to improve the economic condition of the rural people of J&K.

In the Union Territory of J&K, sericulture is considered as an age old vocation. J&K has enormous potential in sericulture or silk production. Commercial sericulture can really empower the farmers of J&K by increasing their income and employment. The raw silk produced in J&K is recognized across the globe. As per researches sericulture is considered as an age old vocation in J&K but in Jammu region it was introduced about 100 years back. Major part of J&K is rain fed and maize, wheat and paddy are the major crops of the region. The size of land holding in J&K is small and horticulture and agroforestry have immense potential in the region. It is difficult for the farmers of J&K to mitigate all their basic needs because of low income from agricultural sector. Some farmers from the particular areas of J&K take sericulture as the subsidiary occupation. In order to promote the sericulture industry in J&K, mulberry trees production needs to be enhanced. There is need to introduce some new varieties of mulberry so that the leaf production can be increase which will facilitate the silkworm production. Generally two rearings of silkworms are conducted one in February-March and another in Aug.-September but if the mulberry leaves are available in sufficient quantity then four rearings are possible in a year. Sericulture is taken as a subsidiary occupation by about 29000 rural households in J&K. Most of these families are economically backward and sericulture plays a crucial role for their livelihood. In order to empower these farm families' productivity and quality of cocoons need to be improved by applying the more scientific package of practices. Some high yielding breeds of silkworms' along with the new varieties of mulberry need to be introduced. Integrated Nutrient Management (INM) is also required for the scientific mulberry cultivation. Farmers need more rearing space and start-up tools for the scientific sericulture. More awareness and capacity building programmes need to be organized for the farmers so that they can be updated with the latest technologies in sericulture. Sericulture is mostly bivoltine in J&K and most of the cocoons are produced during the spring season. Although the autumn crop has been introduced but yield and quality is not excellent. Post-cocoon management also needs to be improved. It has been observed that cocoons reared in J&K are of high quality in Asia. Its fibers can be compared with the best on the world. Inadequacy of mulberry leaves and lack of awareness on scientific technologies are the hurdles for the progress and area expansion of sericulture industry in J&K.

Sericulture is one of the most ancient industries in J&K. J&K silk goods are recognized across the globe because of their colour and quality. There was time when the silk goods were exported from J&K to the Roman, Greek and Persian empires. Mughals were also fond of silk from Kashmir. Many nurseries are established for the production of mulberry saplings which are further distributed among the farmers. Department of Sericulture and State Agriculture Universities (SAUs) of J&K are making strenuous efforts for the area expansion and production of cocoons in J&K. They collaboratively organize the capacity building programmes for the silkworm growers and upgrading their knowledge on sericulture. There are about 0.38 lakh of weavers who are engaged in this industry in J&K. Silk goods from J&K are of high demand in the national and international level. These handloom goods generate employment in J&K. Reasi and Udhampur districts of J&K have huge potential for sericulture. There are certain sericulture pockets in both the districts where the farmers are doing commercial cocoon production. The income from sericulture is supplementing their total income and uplifting their socio-economic status.

J&K is the only state in the country which is at the same altitude in which the leading bivoltine sericulture countries of the world lie. J&K has the natural climate which is congenial for the production of high quality raw silk. Bivoltine sericulture is on the top priority for the Indian sericulture industry. Although India is at second rank in the silk production in the world but the domestic demand is more than that of production. The more demand is mitigated through the import from

China. J&K has huge potential in the production of bivoltine silk and can reduce its import from abroad. In J&K silk industry can be flourished and more income and employment can be generated through it. There is a dire need to give a boost to the sericulture industry in J&K. It can rejuvenate the industrial sector in J&K. The silk carpets from J&K are already are of great demand and with new innovations we can tap the untapped markets. Some more research projects can help in the value addition in the sericulture industry. More and more unemployed youths can be attracted towards this industry by providing them the incentives and subsidies. This sector has been identified as the viable sector for the employment generation. This industry can revolutionize this beautiful Union Territory with a new source of livelihood for the rural and urban people. SKUAST-J, SKUAST-J and Department of Sericulture can play the key role to enhance the quality and quantity of cocoons in J&K. There should be proper mechanism to sale the cocoons and silk products. Strenuous efforts are needed to revive this industry for the benefit of the rural people. There is dire need to promote the silk produced from J&K. The government has launched some small schemes including free plant material and distributing of free silk rearing kits that growers are using during rearing process of cocoons. Sericulture authorities organized cocoon auction markets across J&K where the growers sell their cocoons. In J&K, sericulture is an ideal enterprise for the rural development especially for the weaker sections of the society. This sector has the unique feature of converting family labour into useful income for the family of the farmers. Sericulture is an eco-friendly vocation and has high potential for poverty eradication. It has proved as an ideal tool for women and tribal farmers of J&K. It is an excellent tool for transfer of wealth from affluent section of the society to the poor. A strategic approach is needed to give a boost to the sericulture industry in J&K.

Chapter 8

Saffron Cultivation in J&K

Union Territory of Jammu and Kashmir is a mountainous in which Jammu region is predominantly sub-tropical while Kashmir region is temperate. Agriculture plays the pivotal role in the economy of J&K. Agricultural sector is very vast and it provides employment directly or indirectly to around 70 per cent of J&K population. Agriculture contributes about the significant amount of this UT revenue which signifies the over-dependence of J&K on agriculture. The average size of land holding of J&K is only 0.67 hectare against 1.33 hectares' land holding size on national basis. About 30 per cent of the area of J&K is under cultivation. The agro-diversity of J&K varies from sub-tropical in Jammu region and temperate in Kashmir region. There is need to increase the income and employment in agriculture and allied sectors for the farmers in J&K. This is possible to increase the production and productivity of the crops and to enable the farmers to diversify their crop production. There is need to promote the diversification in agriculture and move towards high value crops such as saffron, vegetables, aromatic and medicinal plants, mushroom, bee-keeping, silk worms, pulses *etc.* J&K is endowed with ample natural resources including soil, water, climatic condition, diversity, topography, rich natural flora *etc.* which are conducive for the cultivation of a wide range of crops.

Saffron (*Crocus sativus*) is a slender, dried, reddish brown, flattened stigma of cultivated saffron plant. It has various names such as Kesar, Zafran, Kang Posh *etc.* Saffron flowers are the symbol of freshness and purity. There are various types of flowers grown in J&K but saffron flowers have their own utility and importance. In the Indian agriculture, saffron cultivation is known as the 'Golden Zest'. Saffron is one of the most important spices in J&K. In India, 5,707 hectares of land comes under saffron cultivation and out of it 4,496 hectares of land is cultivated in J&K. After fruits production saffron cultivation is the second largest activity in J&K. The diverse climatic conditions, fertile soil, sub-tropical, semi-temperate, temperate and arid climatic conditions in J&K offer an immense scope for cultivation and production of different crops. Among the major horticultural products saffron is one of the most important exporting crops and plays an important role in income and employment generation for the rural people of temperate regions of J&K. The main cultivation areas of saffron are Kharewa, Chandhora, Pampore areas. District Pulwama, commonly known as saffron bowl of Kashmir. It is dominant in saffron production followed by Budgam, Srinagar and Kistwar districts. A hundred Kg of fresh saffron flowers yield about three Kg of dehydrated stigmas. Pure saffron consists of only the orange-red stigmas of the saffron plant.

Saffron is very famous in the world due to its fine flavour, colour and medicinal value. It is low volume cash crop. It is estimated that about 49 per cent of saffron produce is exported from J&K. The saffron is one of the oldest commodities of Jammu and Kashmir. It has been observed that every year around 2,128 kilograms of saffron is produced from Pampore and its neighbouring areas. But during last few years the production the saffron cultivation is under threat due to uncertain climatic conditions and the insufficient irrigation facilities in J&K. There is need to enhance the area, production and the productivity of this important spice in J&K as it is a good source of earning foreign exchange for the country as a whole. Saffron cultivation in J&K has its historical roots from Iran where world's largest amount (about70 per cent) of saffron is produced. Saffron has many uses in industries such as food, pharmaceutical, cosmetic and perfumery as well as in the textile dyes. It is very effective in the treatment of various diseases. Saffron stigma has antioxidant activity and thus it prevents the degeneration of cells by free radicals. The components of saffron, crocin and safranal shows role in the suppression of inflammatory pain responses, decreases the number of neutrophils and also possesses strong activity against various bacteria and fungi. It has a protective effect against nephrotoxicity and is cardio-protective. It reduces the fasting blood glucose levels and also exhibits an anti-tumour effect through inactivation or activation of different molecular cascades. It is also used in the manufacture of perfume. It is considered as the panacea for the various diseases of males and females.

J&K has monopoly in the saffron cultivation in India. About 7.3 per cent of world saffron is produced from J&K. India is at third rank saffron production in the world. The matter of concern is that the saffron cultivation has declined over a period of time. However, Saffron production is currently suffering on several counts, especially those relating to productivity as well as post-harvest management. This has resulted in lower production and poor quality. The intense cultivation and monoculture of saffron in saffron grown belts of J&K, together with the continual use of diseased material has resulted in the frequent occurrence of corm rot diseases caused by pathogens such as *Rhizoctonia crocorum, Phoma crocophila, Fusarium moniliforme etc.* a non-sporulating basidiomycetous fungus. Of these diseases, corm rot of saffron caused by *Fusarium oxysporum* and *Fusarium solani* is considered as the most destructive. These infections generally take place through the injury of corms. This disease is quite destructive in saffron and it reduces a considerable

yield of saffron crop every year. In order to control saffron corm rot, the corms need to be treated with a fungicidal solution containing Mancozeb 75WP (0.3 per cent), Carbendazim 50WP (0.1 per cent) for 5-10 minutes and then spread on a cloth and allowed to dry in shade for another 10-15 minutes. Most important fungicides such as Blitox, Indofil M-45, Difolatoan, Folpat, Bavistin and Tecto are very effective to manage this disease.

Lack of research and development and irrigation facilities in the saffron cultivation are the major constraints in saffron cultivation because when the corm do not get sufficient water and rainfall at regular intervals it does not grow properly to give good yield. Farmers lack the awareness to tackle these problems. Also there is existence of the various intermediaries in the marketing in the saffron sale leading to its adulteration. There is need to constitute special policies to strengthen the saffron cultivation in J&K so that the income of the farmers can be enhanced. Special incentives can also be provided to the saffron growers in J&K.

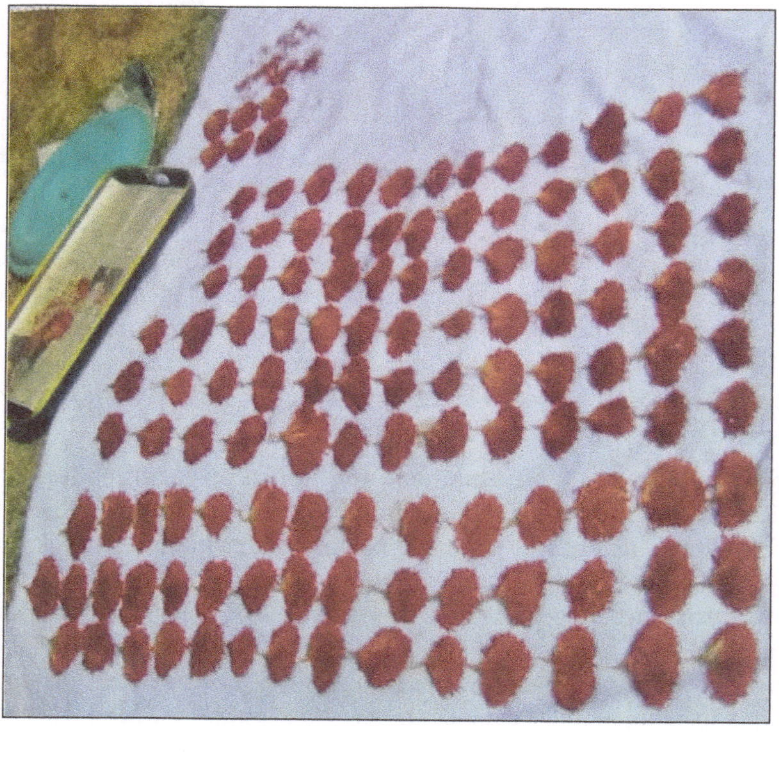

Keeping in view the importance of saffron industry in the J&K, saffron growers can be given more financial support and expertise to protect the future of the saffron cultivation in J&K. Scientific saffron cultivation can generate more employment and income opportunities among the farmers of temperate areas. There is need of quality planting material, co-operative societies and proper trainings for its post-harvest management. In Jammu province some temperate areas have the potential for saffron cultivation. SKUAST-K has also laid Front Line Demonstrations (FLDs) of saffron crop in close co-ordination with the Krishi Vigyan Kendra (KVK), Reasi in temperate areas of Reasi and Udhampur districts. There is need to build strong linkage between the farmers, extension workers and scientists to increase the production and productivity of saffron in J&K.

Chapter 9

Scope of Medicinal and Aromatic Plants in J&K

Union Territory of Jammu and Kashmir is located within the North-Western Himalaya and is endowed with incredible diversity of medicinal plants. Kashmir is often referred to as a 'Terrestrial Paradise' and is well-known across the globe. One of the main features contributing to the worldwide reputation of Jammu and Kashmir is the rich biodiversity that adorns its captivating landscape. The beautiful Union Territory of Jammu and Kashmir harbors a diversity of medicinal plants that have been used in traditional health care systems for thousands of years. This knowledge of health care is transmitted from generation to generation. But from the last few decades it has been eroding due to rapid socio-cultural change. Besides this, medicinal plants are being overexploited at an alarming rate. The traditional uses of medicinal plants in Jammu and Kashmir have not been reviewed properly and if reviewed, focus has been given either on particular groups or on restricted regions of J&K. There is dire need of comprehensive documentation, diseases treatment by the traditional medicinal plants and scientific guidelines for the conservation and management of medicinal plants of Jammu and Kashmir. The main ailments that are treated include dermatological, gastrointestinal, dental, wounds, skeleto-muscular, cough and cold, respiratory, genito-urinary, fever, headache, circulatory and opthalmological. A single plant species may be used to cure various human ailments. The traditional hakims and healers utilize different parts of the plants for the remedial measures of different ailments. However, the use of a particular plant part depends on the plant habit and user's needs. The most frequently used plant parts in the preparation of herbal medicines include leaves, seeds, fruits, roots, rhizomes, whole plants, flowers and bark. Different liquids such as water, sugar, honey, juices, tea, edible oil and milk are mixed with plants or plant parts during the preparation of the medicines for the different ailments. The prepared medicines are mostly administered orally (63 per cent), less frequently dermally (19 per cent) or both orally and dermally (15 per cent). Only 3 per cent is administered through ears or eyes. Keeping depletion of the medicinal plants in view, various Government and Non-Government Organizations (NGOs) are involved in conservation of these plants species. In Jammu and Kashmir, various organizations such as Indian Institute of Integrative Medicine, Jammu (IIIM-Jammu) (formerly known as RRL, Jammu), Sher-e-Kashmir Agriculture University of Science and Technology, Jammu and Srinagar, Centre for Biodiversity Studies, BGSB University, Rajouri, University of Kashmir, Srignar, State Forest Research Institute, J&K, Defense Institute of High Altitude Research, Leh and Shri Mata Vaishno Devi University, Katra and University of Jammu are keenly involved in developing conservation technologies for medicinal plants of J&K. The high demand medicinal plants in Jammu region are Harad (*Terminalia chebula)*, Amla (*Emblica officianalis)*, Ashwagandha (*Withania somnifera)*, Baheda (*Terminalia bellirica)*, Arjun (*Terminalia arjuna)*, Banafsha (*Viola odorata)*, Gilo (*Tinosporia cordifolia)*, Tulsi (*Ocimum sanctum)*, Saffron(*Crocus sativus) etc.* while in Kashmir region the high demand medicinal plants are Gowzaban (*Borage officianalis)*, Asal us soos (*Glycirrhiza glabra)*, Saffron (*Crocus sativus)*, Banafsha (*Voila odorata)*, Behidana (*Cydonia oblonga)*, Gul-e-Zoofa (*Hyssopus officinalis)*, Parshianshan (*Adiantum capillus)*, Kalonji (*Nigella sativa)*, Kansi (*Cichorium intybus)* etc.

As reported by the World Health Organization (WHO) medicinal plants still form the basis of traditional or indigenous health systems and are used by the large part of the populations in most of the developing countries. Medicinal and aromatic plants (MAPs) play a key role in mitigating the demands of the traditional medicine markets which are found both domestically in the producing and in overseas markets. Demand for wide varieties of wild species of medicinal and aromatic plants is increasing with growth in human needs, numbers and commercial trade. It has been observed that some wild species of medicinal plants are being over-exploited. Various agencies are recommending that wild species of medicinal plants should be brought under commercial cultivation. There are various high demand and high commercial value medicinal and aromatic plants in Jammu and Kashmir and there is dire need to assess these herbs in J&K. It has been observed that the trade of raw medicinal plants in the state is not in the streamline with negligible authoritarian control. The high demand species of medicinal and aromatic plants are distinctive with their local availability and the popularity of the herbs is directly proportional to availability of the herb in the area. There is need of expertise in the identification of these herbs.

Since ancient times plants have been used to heal and cure diseases and to improve health and well-being of human beings and animals. Trade in medicinal plants is difficult to estimate accurately because much of the local trade is either unrecorded or poorly classified. Increasing global interest in medicinal plants has also created a sustained and largely underground trade in plant materials, many of which are being collected in an unregulated manner, resulting in indiscriminate

harvest of wild varieties of medicinal plants and serious damage to biodiversity. The annual demand of botanical raw drugs in the country is increasing. This reflects synthesis of data related to estimates of consumption of botanicals by the domestic herbal industry, the rural households and the volume of botanicals recorded for the export. It has been observed that there are various kinds of medicinal plant species found across the three divisions of J&K. While Jammu and Ladakh account for 500 and 880, Kashmir division accounts for around 3000 such species of medicinal plants, respectively. Unscientific and indiscriminate harvesting techniques have pushed most of these species on the verge of extinction. Absence of reliable market information has been an impediment in the management of medicinal plants resources. There is need of strenuous efforts for promoting mass cultivation of natural herbs in J&K. According to the World Health Organization (WHO) the present demand for ethno-medicinal plants is approximately US$14 billion per year. The demand for medicinal plants based raw materials is increasing at the rate of 15 to 25 per cent annually; the demand for ethno-medicinal plants is likely to increase more in future. Excessive anthropogenic pressures have been identified as the main causes of decline in the population and availability of the medicinal plants in the Himalayan region. With increasing demand and renewed global interest in traditional ethno pharmacy, coupled with the increasing preference for natural substances in the healthcare system, the natural stock of medicinal plants of Himalayas is encouraging. Commercial cultivation is crucial for the success of medicinal plants to mitigate the ever-growing demand for medicinal plants as this wealth has been depleting continuously since last two decades in their natural habitat. It has been observed that there are 880 medicinal plants species involved in all India trade. Of this, 48 species are exported and about 42 spices are imported. As per the Ministry of Environment and Forests, Government of India, reveals that there are over 8000 species of medicinal plants grown in India. Despite having only around 2 per cent land of the world, it is blessed with 25 per cent of the biodiversity of the world. The demand for medicinal plants in India to meet both domestic and export markets has been projected to increase it at about 15–16 percent annually, between 2005 and 2010. India exports its medicinal plants products to at least 95 countries. The main destinations were the USA and Europe.

Government of India has launched a scheme known as Jammu and Kashmir Aroma Arogya Gram (JAAG) to encourage the farmers of Jammu and Kashmir to cultivate scientifically tested and commercially beneficial aromatic and medicinal plants in J&K so that their socio-economic status can be uplifted. Under this scheme the government's key R and D body, Council for Scientific and Industrial Research (CSIR) is helping the farmers of J&K with expert guidance and provide them saplings of as many as nine varieties of medicinal and aromatic plants. The plant varieties, selected for Jammu and Kashmir, include five different varieties of lemon grass (*Cymbopogon*), Rose (Rosa), Mint (Mentha), Ashwagandha (*Withania somnifera*) and Phalsa fruit plant. Recommended By Colombia Different CSIR institutions like Indian Institute of Integrative Medicine (IIIM), Jammu, Institute of Himalayan Bio-resource Technology (IHBT), Palampur, Central Institute of Medicinal and Aromatic

Plants (CIMAP) Lucknow, National Botanical Research Institute (NBRI), Lucknow and North Eastern Institute of Science and Technology (NEIST), Jorhat have been associated for the implementation of this scheme in Jammu and Kashmir. The proven agrotechnologies of CSIR institutions have been demonstrated in different areas of J&K. Various medicinal and aromatic plants species such as Lemongrass CKP-25,CPK-F2-38,Rosa grass RRL(J)CN-5,IIIM(J)JK-10,Mint RRL(J) ML4 and Phalsa IIM(J) Shivalik Berry have been developed that can perform well in J&K. It is hoped that entrepreneurship development in this sector will bring prosperity among the farmers of J&K.

Chapter 10

Potential of Organic Farming in Hilly Areas of J&K

Agriculture is the backbone of Jammu and Kashmir and this sector provides employment directly or indirectly to around 70 per cent of the state population. Agriculture contributes about 65 per cent of J&K revenue which signifies the over-dependence of the J&K on agriculture. The Environment Sustainability Index of J&K is worsening and is a matter of grave concern for the scientists and policy makers. There are various factors which are responsible for the environmental catastrophe in J&K. The use of fertilizers and other agrochemicals is on rise in J&K which deteriorate the soil health. Organic farming is a method of farming in which we largely exclude the use of chemical fertilizers, pesticides, herbicides, growth hormones and we maintain the soil fertility by the organic manures, green manures, crop rotation *etc.* It is a unique system of farming which sustains the agro-ecosystem including biodiversity and soil biological activities. According to the National Commission on Farmers organic farming should be the major tool for the second green revolution specially in the hilly areas.

Over the years, the farmers of the hilly Union Territory of Jammu and Kashmir have adopted new agricultural technologies but still J&K is having low productivity of almost all the crops. There is need to develop systematic approach and plan for the development in organic farming in Jammu and Kashmir as J&K has immense potential for organic farming. Several steps have been taken by the government in order to highlight the importance of organic agriculture in the growth of J&K economy. There is need to make rigorous efforts to promote organic farming in J&K has immense scope in this sector. Soil health is deteriorating day- by- day due to indiscriminate use of chemical fertilizers in the soil. In order to sustain the health of soil organic farming is the best option. Organic farming is picking up pace in this

UT and there is need of awareness and trainings of farmers for organic farming. J&K has huge potential for organic farming as a large area in J&K is already under semi-organic cultivation in hilly districts of J&K due to the lack of availability of chemical fertilizers in these areas and the farmers of these areas hardly apply the chemical fertilizers. Basmati rice of R. S. Pura, rajmah of Bhaderwah, potato of Gurez and Machil and red rice of Tangdar, Kupwara, ginger and turmeric of Pouni, Reasi are major exportable organic products in J&K and have the potential to fetch more returns in the market. There is need to explore markets in the country and abroad for saffron, basmati rice, rajmash, spices and various other agricultural produce from J&K.

The farmers of Jammu and Kashmir can tap the opportunity of growing demand of organic products. Organic agriculture is generally environment friendly maintains soil health and enhance biodiversity. Already large numbers of farmers of J&K are growing spices, basmati rice, walnuts, herbs *etc.* organically. There is need to introduce organic farming commercially, scientifically and in a systematized way so that the farmers can get rich dividends and farming can become sustainable. Keeping the importance of organic products there is need to shift from chemical to organic farming. Organic farming system is not new and it is being followed from ancient times. It is a method of farming system which primarily aimed at cultivating the land in such a way so that the soil health can be maintained. There is an immense scope for organic Basmati rice in J&K which is grown on more than 35,000 hectares of land in J&K and has a production of over 88,000 metric tonnes, besides rajmash, spices, saffron, pulses and potato. There is need to provide logistic support for organic farming to the farmers so that they can maximum benefits from it.

Modern crop farming has enhanced the food grain production but it has caused many problems to the environment and human health. Besides, it has contributed to global warming. The imbalanced use of agro-chemicals in soil and on plants is not only damaging the soil bacteria, fungi, actinomycetes *etc.* but has given rise to phenomenon like pest resistance and pest resurgence. Dependence on the external inputs like fertilizers, pesticides, machines *etc.* have increased rural indebtedness and created dependencies. Our farmers have indulged in indiscriminate use of chemical fertilizers and pesticides. In order to increase the agricultural production and control of insect-pests and diseases, our farmers are increasingly depending on agro-chemicals. Now the people are questioning the impact of modern agriculture on environment, economic and social aspects. Many farmers are seeking the alternative practices that would make agriculture more sustainable and productive. Organic farming is the only alternative taking care of all ecological aspects. A natural balance needs to be maintained for sustainability of crop production system. Therefore, recycling of resources through organic farming approach is of paramount importance.

The term 'Organic Farming' was coined by Northbourne (1940).Organic farming is a method of farming system which primarily aimed at cultivating the land and raising crops in such a way,so as to keep the soil in good health by

the use of organic wastes and other biological materials along with beneficial microbes. Nutrients are provided to crops for increased sustainable production in an ecofriendly and pollution free environment. The aim is to create integrated, human, environmentally and economically sustainable production system which maximize reliance on farm derived renewable resources and management of ecological and biological processes and interactions so as to provide acceptable level of crop, livestock and human nutrition, protection from pests and diseases and appropriate return to the human and other resources. Organic farming means holistic production systems which refer earth friendly methods for cultivation and food processing. It differs from other systems by the certification procedures, specific standards and a specific labeling scheme. Organic farming respects the environment by friendly practice of weeds, insect-pests and diseases control. Insecticides, pesticides, chemical fertilizers, growth hormones and antibiotics are not used in organic farming. Ecofriendly techniques are employed to increase the production. Organic farming is based on the integral relationship and we need to understand the relationship between different entities of a farm. The organic farming is based on organic relationship. Organic farming can be perceived as a system which should always be in search of natural alternatives suitable to the local specificity. The organic manures supply the nutrients in the soil, promote microbial activities and improve physical, biological and chemical properties of soils.

Organic farming is a method which recognizes the ecological interconnections between different compartments of cropping. Organic farming is based on certain principles. The organic farmer regards the farm unit as an ecosystem and recognizes that a change in one part of the system can disturb many inter-relationships that exist within the unit. The diversity at farm level creates an ecosystem that has biological checks and balances that help to prevent any one species of insect, disease or weed from becoming a problem. Natural ecosystems are characterized by a continual growth; decay and recycling of nutrients. The farm residual are utilized in the soil to build biological fertility. It is difficult to cycle all the nutrients internally. It has been observed that about 60-90 per cent of the nutrients can be recycled. In advanced natural ecosystem producer and consumers stay together and nutrients loss is completely eliminated. This principle helps to diminish the acute dependence on outside support of nutrients, integration both animal feed production and local settlement. Organic farming is a market demand driven agriculture which aims to fetch more prices. Postharvest crop residues can be utilized to the fullest extent. Organic farming has experienced a dramatic expansion in all the crops due to increase in the awareness in the food security and environment safety. With the awareness on organic farming the area under organic farming is increasing every year. Currently the demand for organic products is more than supply. The market for organic crops growing at a very high rate each year. The reduction in the input cost in organic farming ensures higher economic net returns. The emerging challenge for food security and environment sustainability can be mitigated by the organic farming. Good market demand for organic products will have a positive impact on the socio-economic status of the farmers.

Chapter 11

Talwara: Emerging Organic Village in J&K

Talwara village is around 6km away from Reasi town of Jammu and Kashmir. Baseline survey of village Talwara was done by the KVK experts and it was found that the farmers of the village were interested in the organic farming especially vegetables and spices. Krishi Vigyan Kendra (KVK), Reasi scientists repeatedly approached to the farmers of this village and motivated them for the organic farming. The KVK scientists frequently visited the village and trained the farmers for the organic farming emphasizing on organic vegetables growing. The farmers' clubs'/SHG formation was initiated in the village with the co-ordination of NABARD and ACTECK. The KVK efforts were synergized by the Department of Agriculture and other line departments of the district including NGO. The farmers' clubs were encouraged for the organic vegetables growing in the village. In the beginning the farmers were advised to avoid the agro-chemicals in the vegetables so that that the can be organically grown. The farmers were guided for the establishment of vermicompost units in the village so that the soil nutrients requirements can be fulfilled. In the beginning the farmers faced lot of problems as they were not getting the desired yields of the vegetables. The management of insect-pests and plant diseases was done by using the organic inputs. The quality vegetables seeds and other organic inputs were provided to the farmers. KVK, Reasi expertise were utilized by ACTECH and Department of Agriculture in creating awareness on certification process and other technical guidelines on commercial organic farming. After a long practice on organic farming the certification of their products was done by the Intertech India Ltd. through Agricultural and Processed Food Products Export Development Authority (APEDA). Now the farmers of the village are selling the organic vegetables at the large scale. Their organic vegetables are sold in the local market and also in Jammu at a good price. There are 8 farmers' clubs and one Self Help Group (SHG) of women who are working for the organic vegetables production. KVK, Reasi has also established an Integrated Farming System (IFS) model in the village which attracts the other farmers of the area for the IFS system. This IFS model is producing the organic crops and also has a high tech dairy farm. By this entrepreneurship the farmers of this village generate extra income which develops the confidence among them and gain recognisation in the state.

Establishment of Organic Training Centre at Talwara

Sher-e-Kashmir University of Agricultural Sciences and Technology of Jammu (SKUAST-J) has established an Organic Training Centre at Talwara village of Reasi district which is being sponsored by the Indian Council of Agricultural Research (ICAR), New Delhi under Pt. Deen Dayal Upadhaya Unnat Krishi Shiksha Yojana. Krishi Vigyan Kendra (KVK), Reasi is supervising the centre and upgrading the skill development of farmers especially on organic farming.

Impact of KVK Activities

KVK, Reasi imparted the awareness/trainings programmes/kisan ghosties on organic farming in the village Talwara of Reasi distt.Till date by KVK experts have trained more than 380 participants on organic farming in Reasi district. Before KVK interventions farmers of Talwara village were using their traditional technologies for vegetables growing and they were earning only Rs.73,000/ha of vegetables but with the new scientific technologies of organic vegetables growing they raised their annual income up to Rs. 2,55,000/ha in 2014-15 to Rs. 2,90,300/ha in 2018-19. By growing the organic vegetables, the farmers are fetching more prices. By observing the successful results of organic vegetables growers in Talwara village, the other farmers of the adjoining villages such as Kansi Patta, Kans Brahamna, Seela, Chaiter *etc.* are also showing keen interest for organic vegetables growing in the district. With the establishment of vermicompost units in the village, the farmers are able to fulfil the need of the required soil nutrients in the vegetables growing. With the generation of extra income organic vegetables growers developed confidence to mitigate their basic needs. A self- reliance and entrepreneurial spirit has been developed among the organic vegetables growers in the village. Their new attitude for organic farming has been developed. With the increase in the income of the organic vegetables growers their respect and reconisation have also been increased not only in the village but also in the state. They exhibit their organic farm produce in different farmers' fairs organised by SKUAST-J and allied departments and recognised by the different organisations. Their efforts for organic farming were highly appreciated by Hon'ble ex-DG, ICAR, Dr. S. Ayappan, Hon'ble Union Minister Dr. Jatinder Singh, and Hon'ble J&K Min. Sh. Ajay Nanda during their visit at KVK, Reasi. Former Hon'ble Vice Chancellor of SKUAST-J, Dr. P.K. Sharma of SKUAST-J, Hon'ble Prof. J.P. Sharma also visited this village many times and appreciated the farmers for their efforts towards organic farming. KVK, Reasi is making strenuous efforts for the farmers' welfare through organic farming in the hilly district Reasi of Jammu and Kashmir.

Talwara during Lockdown Period

This year the corona virus pandemic crisis has been coincided with the harvesting of *Rabi* season crops. As the lockdown is going on across the nation, the harvesting of *Rabi* season crops has been initiated by the farmers of Talwara village. KVK, Reasi scientists have suggested some precautionary measures of pandemic to the farmers. Farmers have been suggested to follow the social distancing while harvesting their crops and keep at least 2 metres distance from each other. They have been suggested that they should properly senitize their sickles and other machinery to be used in the harvesting, bundling and threshing of crops. They are appealed to use the masks while harvesting the crops and wash their hands with soap in regular basis. They have also been suggested to drink the clean drinking water and keep it aside at a suitable place in the field. The personal hygiene of the farmers have also been suggested. The farmers have been asked to take consultations and be in touch with the KVK scientists for the further advisories. Organic inputs were also provided to the farmers. Farmers also discussed their marketing problems of their farm produce.

Chapter 12

Krishi Vigyan Kendras (KVKs) for Rural Transformation in J&K

Agriculture is the mainstay of J&K and livelihood of majority of population of this Union Territory revolves around agriculture. J&K is endowed with a wide range of agro-climatic conditions which are conducive for growing different kinds of crops. J&K has inherent agricultural potential. Agricultural development harbingers the overall growth and development of this Union Territory. J&K is a mountainous Union Territory in which about 30 per cent of the area is under cultivation. About 70 per cent area of J&K is under food crops. Judicious use of land is necessary to mitigate the growing needs of the increasing population of this UT by keeping the sustainability of soils, ecosystems and environment in view. The average size of land holding of the J&K is only 0.54 hectare as against 1.33 hectares' land holding size on national basis. The agro-diversity of J&K varies from sub-tropical in Jammu region and temperate in Kashmir region. The average annual rainfall of these two regions is 1069mm and 660mm. The average temperatures of these two regions are 24.5 and 13.3 Centigrade respectively. Maize, wheat, paddy, pulses, oilseeds, potato and okra are the main crops of J&K. The farmers of J&K are now shifting towards high value crops such as flowers, vegetables, aromatic and medicinal plants, mushrooms *etc.* In some pockets of J&K Basmati rice, Rajmash, saffron, honey beekeeping *etc.* are also cultivated. Major part of J&K suffers from lack of irrigation facilities, remoteness, lack of transportation facilities, regular soil erosion, inaccessibility of quality agro-inputs *etc.* Lack of storage facilities, small size of land holdings, lack of agri-entrepreneurships, inaccessible terrains, lack of market networks, lack of farm mechanisation *etc.* are the other constraints for the farmers. These constraints hamper to increase the agricultural production in J&K and it further effects the income and employment generation in the rural areas. Although the production, productivity and area under different crops have increased over the years but still the development is slow. Cropping intensity of Jammu region is 176 per cent whereas in Kashmir region it is 123 per cent. The agriculture being less remunerative profession is unable to attract the rural youths of J&K towards agriculture. They are migrating towards urban areas to earn their livelihood. There is dire need to make some strenuous efforts to make agriculture a more remunerative profession.

There is urgent need to bridge the gap between production and consumption of food grains. There is also need to increase the income and employment in agriculture and allied sectors for the farmers of J&K. It has been observed that there is an immense potential to increase the production and productivity of different crops grown in J&K. There is need to enable the farmers to diversify their crop production by adopting the modern technologies in agriculture and establishing the infrastructure for the farm production. Diversification in agriculture needs to be promoted. This beautiful UT is blessed with ample natural resources including soil, water, climatic condition, diversity, topography, rich natural flora *etc.* which are conducive for the cultivation of a wide range of crops. About 58 per cent area under agriculture is rainfed and remaining 42 per cent is irrigated in J&K.

The Krishi Vigyan Kendras (KVKs) are considered as the resource centres for the agricultural technologies. They are also known as Farm Science Centers, a gross root level scheme which was designed by the Indian Council of Agricultural Research (ICAR) in the country. In India the first Krishi Vigyan Kendra (KVK) was established in 1974 at Pudducherry under Tamil Nadu Agricultural University, Coimbatore by the ICAR, New Delhi. Presently, the ICAR has established 713 KVKs across the nation. In Jammu and Kashmir, the ICAR has established 17 KVKs under two agricultural universities SKUAST-Jammu, SKUAST-Kashmir and one KVK under Central Institute of Temperate Horticulture (CITH).SKUAST-J has 7 KVKs in Jammu, Samba, Kathua, Reasi, Rajouri, Poonch and Doda districts respectively while SKUAST-K has 10 KVKs in Anantnag, Bandipora, Budgam, Gandarbal, Pulwama, Kulgam, Kupwara, Shopian, Srinagar and Baramulla respectively. Kargil-I, Kargil-II (Zanskar), Leh-I and Leh-II are the four KVKs established in Ladhakh Union Territory.All these KVKs are working under the ATARI (Agricultural Technology Application Research Institute (ATARI), Zone-1 established at Punjab Agricultural University, Ludhiana. Three KVKs in Udhampur, Ramban and Kishtwar of Jammu region are still to be established. These KVKs are catering the needs of farmers of their respective districts and are providing expertise to the farmers. These KVKs are endeavouring to increase the income and employment among the rural areas of J&K. These KVKs have done exemplary work for the farmers of J&K and have established many success stories at national level. Farmers of these different districts of J&K are considering these KVKs as the knowledge hubs of agricultural technologies and every day they are seeking agricultural information/knowledge from the KVK scientists. KVKs are also providing quality seeds and planting material to the farmers. The demonstration units established at the KVKs farms attract the farmers towards them. These KVKs are working on 'learning by doing' pattern

as the farmers are trained for different enterprises. These KVKs are the nerves of the both agricultural universities *i.e.,* SKUAST-J and SKUAST-K. These KVKs are mitigating the scientific agricultural needs of the farmers of Jammu and Kashmir and assisting the farmers to enhance their income and employment through technological interventions. These KVKs are playing a pivotal role in transformation in rural areas of J&K by upgrading the rural people with the latest agricultural technologies. These KVKs are immensely playing a crucial role in farmers' prosperity. The KVKs of J&K have proved their worth to mitigate the agricultural needs of the farmers. They are doing the commendable work in J&K by transferring the latest agricultural knowledge to the farmers. These KVKs are able to empower the farmers/rural youths/women farmers through need-based farmers/vocational/ skill development trainings. They are promoting the technology-led farming. They are stressing on maximum use of ICTs during COVID-19 so as to disseminate the agricultural latest knowledge even in far-flung areas. They are bridging the gap between the laboratories and farmland. These KVKs are working in collaboration with the Panchayati Rural Institutions, different departments, NABARD, NGOs, NYK *etc.* These KVKs are able to disseminate the agricultural information from lab to land. Follow-up of different capacity building programmes is always done by the multidisciplinary team of experts. For dissemination of agricultural information in the mass scale print and electronic media always support the KVKs. It has been observed that with the endevours of KVKs, the adoption rate of new agricultural technologies among the farmers is increasing substantially in J&K.

Chapter 13

Commercial Floriculture: An Emerging Industry in J&K

Union Territory of Jammu and Kashmir is located between 32.17 and 37.06 North latitude and 73.2 and 80.36 East longitude in the Himalayan region. This Union Territory is blessed with immense natural beauty. J&K is endowed with ample natural resources including soil, water, climatic condition, diversity, topography, rich natural flora *etc.* which are conducive for the cultivation of a wide range of flowers. It is considered as the paradise on the earth because of its natural glamour. J&K has undergone many changes through the ages, influenced by the different kinds of cultures, religions and rulers during various periods of history. It is this scenic beauty and bountiful nature that lured the Mughals and they were impressed by the beauty of the state and they established many gardens in the valley along their main sites and journey routes. Beautiful gardens in the state are being enjoyed by the tourists across the globe. Keeping the demand of flowers in view there is need to increase the flower production in the state. With the strenuous efforts of Government organizations awareness on floriculture industry has been created among the farmers and some farmers especially from the young generation are showing keen interest in floriculture. A good number of progressive farmers in almost all the districts of J&K have been trading flowers in the local and outside markets. In 1990 the area under floriculture in J&K was only 80 ha and now it has increased around 350 ha with an annual turnover of about 1350 lakhs.

Flowers have their importance because of their aesthetic value in their use in decorating homes, in social events such as marriages, social gathering and funerals, in religious rituals *etc.* Main floriculture crops such as marigold, gladiolus, lilies, tulips, carnations, roses and gerberas are being cultivated in the state to generate additional income. Both the agricultural universities in J&K (SKUAST-J and SKUAST-K) and Department of Floriculture provide technical assistance to the floriculture farmers and impart training to start the entrepreneurship in the floriculture. Technical guidelines on cut flowers, seeds, bulbs and ornamental plants, introduction of new technologies like high-tech poly structures *etc.* are provided to the farmers by these organizations. These organizations also demonstrate some advance floriculture farms to the farmers through exposure visits.

Flowers play an important role in our life right from birth to death. Floriculture industry has been emerged as an important agri-business in Jammu and Kashmir. In view of fast development of technology, to grow the good quality of flowers, their plucking, preservation and transportation this activity has assumed the status as an important industry in J&K. In J&K this industry has started flourishing on scientific basis. J&K has good soil and conducive weather conditions for flowers cultivation. Recently the floriculture industry is rapidly increasing in J&K. With the advance research in floriculture the new varieties of flowers have been developed. Jammu and Kashmir has immense scope of floriculture as the state is blessed with some world famous religious shrines and tourists across the world visit to these spots. The floriculture is not getting the priority as it deserves in J&K. It is an intensive type of farming. In floriculture income per hectare is much higher than many other agricultural sectors if it is done in a scientific way. We should not forget the aesthetic value of flowers in our daily life. Colourful flowers in the house compounds and in the parks really enhance the beauty of our homes, villages, towns and cities. Flowers bring the happiness in our life and boost the sagging spirits. Commercially floriculture increases the income of our farmers. If the government makes the cooperative societies for the flowers growers then the menace of middlemen can be eliminated. The cosmetic industries has also been encouraged which use wide varieties of floricultural products. The essential oils from different flowers such as rose, jasmine and tuberose used in the manufacture of cosmetics have very high price. The business of nursery in the floriculture industry can earn a lot of money for the rural people. One vital step in this industry is needed to certify the seeds and standardizing the floriculture nursery stocks.

J&K has lot of scope of floriculture industry. It has been observed that flowers have high demand within and outside of J&K. We have good soil and weather conditions in J&K which are conducive for the growth of flowers. All these factors indicate that there is immense scope of floriculture in J&K. Now the farmers are becoming aware about this rising industry in J&K. If we look our economy, our agriculture, horticulture and floriculture have attracted the attention of big investors. Floriculture requires higher level of information, economics and technology, the consultancy services in preparing the project reports, to prepare the market survey report, evaluation studies *etc.* National Horticulture Board under the Ministry of Agriculture has introduced many new schemes to encourage the floriculture industry. There is need to promote the cold storage facilities for the floriculture industry. For high quality of floriculture we primarily need the high quality plants which can give quality flowers. For this purpose we need to develop

quality nurseries in the villages which will help to create the employment in the rural areas. Market is another aspect in floriculture industry. The trade of flowers and their products are fastly increasing annually. Certain areas have been specified which are considered conducive for the certain varieties of flowers. Marigold be utilized to produce more and more seeds. J&K has favourable weather conditions for floriculture and we have a good opportunity for exporting the flowers. To produce the quality flowers we need to grow them under the controlled conditions. For this purpose we can use the green houses/polyhouses. Green house is a framed structure made up of either cane wood or steel. The transparent and translucent plastic materials are used. In such green houses the temperature, humidity and carbon dioxide are fully controlled which creates the conducive atmosphere for the growth of the plants. The advantage of green house is that only the congenial radiation which is required for the plant is passed through the sheet while the harmful radiation is trapped. Those farmers who are interested to grow the flowers under green houses should consult with the qualified floriculture experts having good knowledge of plant engineering, production and market aspects also. We need to have more high tech floriculture projects in operation all over the state. If we look at the potential of floriculture industry in the state, it can be said that there is need of skilled and unskilled persons right from planting the plants, plucking the flowers, their storage and their transportation in different local, national and international markets so that ultimately flowers can be reached to the consumers. The process involved in the floriculture industry requires the latest technologies. For this purpose the farmers who are interested to flourish this industry should be trained and necessary infrastructure facilities should be provided to them. Looking at the scope of floriculture industry special trainings are imparted at the village level by the government organizations which help the farmers to start the floriculture projects. These kinds of projects will create employment for the rural people and ultimately the economic status of the farmers of J&K can be improved.

Chapter 14

Kanjali: Emerging Floriculture Village in J&K

Jammu and Kashmir is a mountainous UT in which Jammu region is predominantly sub-tropical while Kashmir and Ladakh regions are temperate. This UT is blessed with an immense natural beauty and is located between 32.17 and 37.06 North latitude and 73.2 and 80.36 East longitude in the Himalayan region. J&K is characterized by benevolent people, scenic natural beauty, natural waterfalls, folded mountains, alpine pastures, lush green forests, rich wildlife, snow clad fields, carpet green turfs, gushing fountains, lofty skying grounds, charming gardens, cool pollution free fresh air, fresh cool water, shimmering lakes, apple orchards *etc.* Beautiful gardens in J&K are being enjoyed by the tourists across the globe. Keeping the demand of flowers in view there is need to increase the flower production in the state. With the strenuous efforts of government organizations awareness on floriculture industry has been created among the farmers and some farmers especially from the young generation are showing keen interest in floriculture sector. A good number of progressive farmers in almost all the districts of J&K have been trading flowers in the local and outside markets. In 1990 the area under floriculture in J&K was only 80 ha and now it has crossed 350 ha with an annual turnover of more than 1350 lakhs. There is dire need to increase the income and employment of the farmers of the hilly UT of J&K. There is possibility to increase the production and productivity of floricultural crops and to enable the farmers to diversify their crops production by adopting the modern technologies and establishing the infrastructure for the farm production in hilly areas of J&K. There is urgent need to promote the commercial cultivation of marigold in J&K as there is immense potential of marigold cultivation in J&K. It is matter of grave concern that the interest of younger generation in the state is dwindling towards agriculture.

Reasi is a hilly district of Jammu and Kashmir which is located around 72 km away from Jammu and is at 1528 meter above mean sea level. Its population is 3, 14,714 as per 2011 census. It was carved from Udhampur district on 1st April, 2007. This district is surrounded by Udhampur, Rajouri, Jammu, Ramban and Shopian districts. This district of Jammu and Kashmir is world famous because of Shri Mata Vaishno Devi Shrine. Every year millions of devotees visit this holy shrine to pay homage. The Chenab River passes through this district which is another centre of attraction in the district. Salal Power Project has been constructed over this river and this generates around 690 MW power. Reasi district of Jammu and Kashmir is an abode of Mata Vaishno Devi and blessed with vast biodiversity. Internationally renowned places like Mata Vaishno Devi, Katra along with some historical places like Shiv Khori Shrine, Dera Baba Banda Bahadur, Baba Aghar Jitto Ji, Siarh Baba, Nao Devian, Dhansar Baba, Ziarat Baji Ismail *etc.* come under Reasi district and tourists from all over the world used to come in the district throughout the year. Being the hub of religious spots, Reasi district is having immense scope in floriculture. The prestigious Salal Hydro Electric Project located at Jyotipuram, Shri Mata Vaishno Devi University, Katra, Krishi Vigyan Kendra, Tanda, and Cancer Hospital, Katra are centres of attraction in the district. The proposed train to Kashmir passes through this district and makes it an important place in the national atlas. The district is watershed of the River Chenab and its major tributaries are Banganga, Anji. Ans, Rudd, Plassu and Pai.

Most of the farmers of Reasi district are small and marginal and crops are mostly cultivated under rain fed (about 94 per cent) conditions. Major crops grown in this area are maize, wheat, paddy, mash and potato. Farmers of this district are having very limited resources and mostly they adopt the traditional system of agriculture. The yield of all the crops is low as compared to the national and state average yields. The adoption percentage of modern agricultural technologies is low resulting in the low socio-economic status of the farmers. There are various insect-pests and disease problems in the district. Boars and monkeys too damage the crops to a large extent in this particular district causing a huge loss to the farmers.

Kanjali village is about 36 km away from world famous religious place Katra in Reasi district of Jammu and Kashmir. The village is around 6 km. away from Krishi Vigyan Kendra (KVK), Reasi. This village is situated in the remote area of Reasi district and is lacking the road and communication connectivity. There are around 150 families in the village. The major occupation of the people is agriculture and there is a natural source of irrigation in the village where the marigold cultivation can be done. There are two Govt. schools, three Angadwadi centres and one Panchayat Ghar in the village. There are many progressive floriculturists in the village who are producing the marigold flowers at a commercial level. The village is having around 150 hectares of land. Initially the villagers were growing maize, wheat and few local varieties of vegetables as their forefathers were doing. They were using the traditional agricultural technologies in their fields and monkeys and wild boars were the great threats in their fields. All the agricultural produce was consumed by their families with very little surplus to sell. Their farming system was deteriorating and they were unable to mitigate the basic needs of their families. Some farmers of the village were on the threshold to leave the agriculture profession.

Before the KVK interventions the farmers of Kanjali village were doing the unscientific agriculture on a small scale. They were endouvering to increase their farm income. They were using the local resources and were growing the maize, wheat, fruits, vegetables, marigold *etc.* with the traditional methods resulting in the poor yield of the crops. They were lacking the scientific knowledge in agriculture. After establishing the Krishi Vigyan Kendra (KVK), Reasi near their village, they came in contact with the KVK, Reasi scientists. KVK, Reasi became a ray of hope for these farmers. KVK scientists started regularly visiting their farms and developed a good rapport with the farmers of this village. A benchmark study was conducted by the KVK scientists in the village. The farmers were guided and motivated for the scientific marigold cultivation. They were encouraged for commercial marigold cultivation. KVK, Reasi organised the trainings/awareness programmes/kisan ghosties/field days/exposure visits/farmers fairs *etc.* on commercial floriculture at Kanjali village of Reasi distt of J&K. KVK, Reasi scientists laid Front Line Demonstrations (FLDs) and On Farm Trials (OFTs) on marigold crop in the village. The technical guidelines on marigold cultivation were provided to the farmers. The farmers were fascinated by observing the results of FLDs and OFTs on marigold crop and they showed keen interest in the adoption of the technologies for marigold crop. The technical literature on commercial floriculture was provided to the farmers of the village and special video shows on scientific floriculture were organised for them. The KVK work was synergised by the Department of Floriculture. Scientific cultivation of marigold changed their life. KVK, Reasi also introduced some new varieties of marigold in the area gave some scientific recommendations on commercial marigold farming. Marigold cultivation helped to overcome the problem of monkeys menace in the village as the crop is not damaged by the monkeys.

Before KVK interventions farmers of Kanjali village were growing only maize, wheat and few vegetables by using the traditional technologies and they were lacking the technical skills and knowledge on agriculture. There was heavy infestation of diseases and insects-pests and farmers were earning very less as their crops productivity was low but with the commercial marigold cultivation they raised their annual income substantially. By growing the marigold crop, the farmers are generating more income as compared to their traditional farming.

By observing the successful results of marigold crop in Kanjali village, the other farmers of the district villages such as Rajwal, Bhabber, Bhakta, Sool, Kansi Patta *etc.* are also showing keen interest in marigold cultivation. With the introduction of scientific technologies for marigold cultivation the farmers are able to generate extra income and employment. With the generation of extra income, the farmers developed confidence to mitigate their basic problems. A self- reliance and entrepreneurial spirit has been developed among the farmers. Their new attitude towards scientific marigold cultivation has been developed. Kanjali farmers' success in commercial floriculture is inspiring the other farmers of the area and they are shifting towards the commercial floriculture in a scientific way for better economic returns. The other farmers of the area take opinion from them about the scientific marigold cultivation practices. With the scientific marigold cultivation, the socio-economic status of the farmers has been improved.

KVK in the form of farmer's trainings, vocational trainings, Front Line Demonstrations (FLDs), On Farm Trials (OFTs), farmers-scientists interaction, farmers' exposure visit, field days, Kisan Ghosties, campaigns, dissemination of production technologies through radio, TV, extension literature *etc.* have led to start the commercial floriculture in Kanjali village. Gradually many farmers of the area are becoming interested in scientific marigold cultivation. They have developed a good rapport with the KVK scientists and always participate in the agricultural extension activities conducted by the Krishi Vigyan Kendra (KVK), Reasi. Year- by-year they have started to generate more income and employment by growing the marigold crop at commercial level. Number of farmers including Sh.Shail Singh, a progressive floriculturist in the village was highly impressed with the technical guidelines given by the KVK scientists and presently he is guiding the other farmers of the area on commercial marigold cultivation. He grows marigold in around 60 Kanals of the area. He has established one vermicompost unit at his home. There are many other farmers in the village who have become entrepreneurs in floriculture like Sh. Shail Singh. They used to sale their marigold flowers at world famous place Katra, Udhampur and Jammu. Now the village farmers have developed confidence in themselves and they are having lot of credibility on KVK scientists as from to time they provide technical advice on commercial marigold cultivation. The village farmers have become an example of success and motivating factor for other farmers in adjoining areas. KVK scientists also motivated other farmers of the village for the commercial floriculture as it can change their socio-economic status. Farmers were facilitated to develop the marketing channels to sale their farm produce in local markets, Katra and even in Jammu. Katra has an immense potential for the marketing of marigold flowers as the tourists from across the globe visit this holy spot and pay homage to the Vaishno Devi Shrine.

Sh. Shail Singh, a progressive farmer of this village was felicitated by the Hon'ble Union Minister of India, Sh. Jatinder Singh in a farmer fair organised by the Krishi Vigyan Kendra (KVK), Reasi of Sher-e-Kashmir University of Agricultural Sciences and Technology, Jammu (SKUAST-J) as a progressive floriculturist in the year 2016-17. He was also appreciated by the Hon'ble Ex-Director General of the ICAR, Dr.S. Ayappan and Hon'ble Vice Chancellor of SKUAST-Jammu Dr.P.K.Sharma for his efforts for commercial floriculture during their visit at KVK, Reasi. Smt. Maya Devi, a woman farmer from the same village was also appreciated by the Hon'ble Ex-Director General of the ICAR, Dr.S. Ayappan for her efforts for commercial floriculture during his visit at KVK, Reasi. Efforts of the Krishi Vigyan Kendra (KVK),

Reasi scientists and progressive farmers of Kanjali village have given a new shape to the marigold farming and the quantity and quality of marigold crop have been increased with new interventions. By observing the results of scientific marigold cultivation at Kanjali village the other farmers of the 'adjoining villages are also attracted towards scientific marigold cultivation. With the increase in the income of the farmers by growing the marigold crop their respect and reconisation have also been increased in the society. They have built their Pucca houses and their children education status is also improving year after year. They exhibit their marigold produce in different farmers' fairs organised by SKUAST-J and allied departments and recognised by the different organisations. Their efforts for commercial farming were highly appreciated by Hon'ble ex-DG, ICAR, Dr. S. Ayappan, Hon'ble Union Minister, Dr. Jatinder Singh, Hon'ble Vice Chancellor of SKUAST-J Dr. P.K. Sharma and Hon'ble J&K Min., Sh. Ajay Nanda during the farmers' fairs and other extension activities conducted by KVK, Reasi. KVK, Reasi is making strenuous efforts for the farmers' welfare through commercial floriculture in the hilly district Reasi of Jammu and Kashmir.

Chapter 15

Kishtwar: A Unique Land of Saffron in J&K

Jammu and Kashmir is a mountainous Union Territory in which Jammu region is predominantly sub-tropical while Kashmir regions are temperate. Agriculture plays the pivotal role in the economy of this beautiful Union Territory. Agricultural sector is very vast and it provides employment directly or indirectly to around 70 per cent of J&K population. Agriculture contributes about the significant amount of UT revenue which signifies the over-dependence of the UT on agriculture. The average size of land holding of the UT is only 0.67 hectare against 1.33 hectares' land holding size on national basis. About 30 per cent of the area of the UT is under cultivation. The agro-diversity of the UT varies from sub-tropical in Jammu region and temperate in Kashmir region. The average annual rainfall of these two regions is 1069mm, and 660mm respectively. The average temperatures of these two regions are 24.5 and 13.3 Centigrade respectively. There is need to increase the income and employment in agriculture and allied sectors for the farmers of J&K. This is possible by increasing the production and productivity of the crops and to enable the farmers to diversify their crop production. There is also need to promote the diversification in agriculture and move towards high value crops such as saffron, vegetables, aromatic and medicinal plants, mushroom, beekeeping, silk worms *etc.* J&K is endowed with ample natural resources including soil, water, climatic conditions, diversity, topography, rich natural flora *etc.* which are conducive for the cultivation of a wide range of crops.

Saffron (*Crocus sativus*) is considered as the highest priced spice in the world. Saffron is a slender, dried, reddish brown, flattened stigma of cultivated saffron plant. It has various names such as Kesar, Zafran, Kang Posh *etc.* Saffron flowers are the symbol of freshness and purity. There are various types of flowers grown in J&K but saffron flowers have their own utility and importance. In the Indian agriculture, saffron cultivation is also known as the 'Golden Zest'. Saffron is one of the most important spices in J&K. J&K has monopoly in the saffron cultivation in India. In India, 5,707 hectares of land comes under saffron cultivation and out of it 4,496 hectares of land is cultivated in J&K. After fruits production saffron cultivation is the second largest activity in the state. J&K has monopoly in the saffron cultivation in India. About 7.3 per cent of world saffron is produced from J&K. India is at third rank in saffron production in the world. The diverse climatic conditions, fertile soil, sub-tropical, semi-temperate, temperate and arid climatic conditions in J&K offer an immense scope for cultivation and production of different crops. Among the major horticultural products saffron is one of the most important exporting crops and plays an important role in income and employment generation for the rural people of temperate regions of the UT. The main cultivation areas of saffron are Kharewa, Chandhora, Kishtwar and Pampore in J&K. District Pulwama, commonly known as saffron bowl of Kashmir. It is dominant in saffron production followed by Budgam, Srinagar and Kishtwar districts. A hundred Kg of fresh saffron flowers yield about three Kg of dehydrated stigmas. Pure saffron consists of only the orange-red stigmas of the saffron plant. It is estimated that about 49 per cent of saffron produce is exported from J&K state. Saffron cultivation in J&K has its historical roots from Iran where world's largest amount (about 70 per cent) of saffron is produced.

Kishtwar district is situated at an altitude of 5351 ft. from the MSL with a temperate climate. The soils of Kishtwar district are appropriate for the saffron cultivation on the basis of soil texture, soil structure, soil profile, PH value and chemical characteristics. Kishtwar is a unique land of saffron. It is well known for quality saffron production. The district is blessed with lofty mountains and sloppy hills around which bestow upon its grandeur and fascinating look for the onlookers.In Jammu province, Kishtwar is the only Saffron producing district. Kishtwar saffron is very famous due to its fine flavour, colour and medicinal value. It is a very popular crop in the region and is considered as the costliest crop for the farmers of Kishtwar. It is low volume cash crop. The saffron is one of the oldest commodities of Kishtwar district. The main marketing channels are brokers/local traders/cooperative societies *etc.* Villages namely Thakraie, Kuchhal, Kuchhan, Bunjwah, Inderwal/Chatru, Palmar, Trigam, Poochal, Dool, Padyarna, Na-gseni, Paddar (Ligri and Tata Pani) and some villages of Drabshala and Dachhan blocks are known for commercial cultivation of saffron in Kishtwar district. About 202 hectares of land in Kishtwar district is occupied by the saffron. About 5-6 quintals of saffron is produced annually from these areas. It takes about 4, 50,000 stigmas or 1,50,000 flowers to produce 1 Kg of saffron spice. The flowering stage of saffron starts from Oct./Nov. and continues for three weeks. Each flower lives for about 48 hours. Traditionally harvesting is carried out by hand. But during the last few years the production the saffron cultivation is under threat due to uncertain climatic conditions and the insufficient irrigation facilities in the district. There is need to enhance the area, production and the productivity of this important spice in the

district as it is an excellent source of earning for the farmers. Quality research and irrigation facilities are needed in the saffron cultivation in the district. There is also need to evaluate the possibility of 'organic farming' of saffron in the district for improving farmers' income opportunities.

Saffron has many uses in industries such as food, pharmaceutical, cosmetic and perfumery as well as in the textile dyes. It is very effective in the treatment of various diseases. Saffron stigma has antioxidant activity and thus it prevents the degeneration of cells by free radicals. The components of saffron, crocin and safranal shows role in the suppression of inflammatory pain responses, decreases the number of neutrophils and also possesses strong activity against various bacteria and fungi. It has a protective effect against nephrotoxicity and is cardio-protective. It reduces the fasting blood glucose levels and also exhibits an anti-tumour effect through inactivation or activation of different molecular cascades. It is also used in the manufacture of perfume. It is considered as the panacea for the various diseases of males and females.

There is need to constitute special policies to strengthen the saffron cultivation in Kishtwar district so that the income of the farmers can be enhanced. Special incentives should be provided to the saffron growers in the district. Farmers of Kishtwar district have positive attitude towards saffron crop cultivation. Keeping in view the importance of saffron industry in Kishtwar, saffron growers need more financial support and expertise to protect the future of the saffron cultivation in the district and they need more technical guidelines. There is need of quality planting material, co-operative societies and proper trainings for its post-harvest management. There is need to build strong linkage between the farmers, extension workers and scientists to increase the production and productivity of saffron in Kishtwar district. There is ample scope for maximizing profitability of this crop for Kishtwar saffron growers, provided that sincere efforts are made. Initiatives are needed by adopting strict quality control measures, mechanizing production and introducing marketing interventions. Adoption of novel scientific technologies, including biotechnology can go a long way to reduce the costs of saffron production in the area.

Chapter 16

Lemon Grass Farming: A Viable Crop for the Farmers of J&K

Monkeys have left a substantial dent in the agricultural production of Union Territory of J&K. Their numbering is increasing day-by-day and endevours are being made by the different organisations to contain the assaults caused by the monkeys. Monkeys are wreaking havoc in certain pockets of J&K affecting the income of the farmers. Monkeys menace has compelled many farmers to keep their cultivable land fallow resulting in the net loss to the farmers. Lack of irrigation facilities and low yield of crops are other problems for the farmers of the area. More pronounced damage caused by these animals in lands adjacent to forest areas is due to food and water shortage in the forests. The major crops of the farmers from certain pockets in J&K are severely affected by the monkeys. Many endevours are made to eradicate this problem but it is increasing day-by-day. Not just village residents, city dwellers are also struggling to cope with monkey menace. Jammu region of Union Territory of J&K is also struggling to manage this menace and many times we observe monkeys attacks on the people. Turmeric, ginger, marigold, medicinal and aromatic plants cultivation *etc.* are suggested by the farmers by the agricultural scientists to minimize the effects of monkey menace. Painthal belt of Reasi District in J&K is heavily infested with the monkey menace. Many fields of the farmers are not cultivated due to monkeys menace. Lack of irrigation facilities, low yield of crops and lack of high milk yielding animals are other problems for the farmers of the area. This led many farmers to migrate in the urban areas to earn their livelihood. In 2016-17 KVK, Reasi with the close co-ordination of CSIR-Central Institute of Medicinal and Aromatic Plants (CIMAP)-Lucknow and Indian Institute of Integrative Medicines (IIIM)-Jammu introduced lemongrass farming in Reasi district of J&K. Now a distillation unit for the extraction of lemon grass oil has been established at Sirah village. Farmers of this area are processing their lemon grass in this unit. It is saving the transporting charges of the farmers as before the establishment of this unit farmers were transporting their lemon grass at IIIM-Jammu and processing their lemon grass crop in its distillation unit. It is worthwhile to mention here that lemongrass is not affected by the monkeys and other animals and farmers can fetch more money by growing it commercially. Many new farmers of the area are showing keen interest in its cultivation. Lemongrass has immense potential in Jammu region of J&K and can be grown by the farmers in their fallow land. In the beginning KVK, Reasi facilitated the farmers in extraction which was done by IIIM-Jammu.Technical guidelines of lemon grass farming are continuously provided by the KVK, Reasi. KVK is facilitating the lemon grass farmers with the co-ordination of district administration and Deptt. of Agriculture. Sh.Tilak Raj, a progressive farmer of Sirah village in Painthal block of Reasi district has grown more than lemongrass in more than 1 acre of land. He is a diversified farmer and generating extra income by growing lemongrass. Presently he is having around 9 lakhs of lemongrass slips at his field which can help him to generate more money. Meanwhile his lemon oil is of very good quality and has a good demand. According to him, lemon grass farming is supporting him to generate more income. By observing the success of lemongrass farming of Sh.Tilak Raj and some other farmers of the area, many new farmers are coming forward for lemongrass farming as the area is severely affected by the monkey menace. Sh.Tilak Raj himself also wants to increase his area under lemongrass farming. Lemongrass oil has tremendous scope in Reasi district as the tourists across the nation visit Katra throughout the year and there are numerous hotels and restaurants where it can be used as the room freshener. It can also be used as the antibacterial and in making lemon tea.

Lemon grass (*Cymbopogan flexuosus*) is a tall perennial plant with thin and long leaves. It is indigenous to India and other parts of Asia. It is aromatic tall sedge of family Poaceae which is grown in many parts of tropical and sub-tropical South East Asia and Africa. In India, it is mainly cultivated in Maharashtra, Kerala, Karnataka and Tamil Nadu states, in some parts of Uttar Pradesh and Uttarakhand besides foot-hills of Arunachal Pradesh and Sikkim. Jammu region of J&K has the potential for its cultivation. If proper package practices and marketing channels are developed for the farmers then their income and employment can be enhanced by the lemon grass farming. Krishi Vigyan Kendra (KVK), Reasi with the close co-ordination of CIMAP-Lucknow and IIIM-Jammu have introduced its cultivation in lower belts of Reasi district of J&K. These belts area severely affected by the monkeys menace and lemongrass has immense potential in the area. District Administration, Reasi and Deptt. of Agriculture, Reasi are facilitating the farmers in the area expansion and marketing of lemon grass oil. Value addition of lemon grass oil is also under process.

Lemongrass is one of the commercially cultivated aromatic crops in India. India is the largest producer of lemon grass and about 80 per cent of the produce is being exported. The essential oil is being traditionally exported to West Europe, U.S.A. and Japan. It has lot of medicinal properties and health benefits. It can be used in regular tea consumption for the best aromatic flavor. Most of the species

Sirah
Lemon Grass
FRAGRANCE OIL

of lemon grass are native to South Asia, South-east Asia and Australia. The so called East Indian lemon grass (*Cymbopogon flexuosus), also known as Malabar or Cochin grass is nati*ve to India, Sri Lanka, Burma and Thailand; for the related West Indian lemon grass (*C. citratus*), a Malesian origin is generally assumed. Both the species are today cultivated throughout tropical Asia. Its culm is stout, erect, up to 1.8 meter high. Leaves are long, green, linear tapering upwards and along the margins; ligule very short; sheaths terete. It is a short day plant and produce profuse flowering. The inflorescence is a long spike about one metre in length. Flowers borne on decompound spatheate, panicles 30 to over 60 cm long. Presently it is grown in about 3,000 ha area in India, largely in states of Kerala, Karnataka, U.P., Uttrakhand and Assam. Now it is also grown commercially in Jammu region also with very good results.

The oil is distilled from its leaves and flowering tops of lemon grass. The oil has strong lemon-like fragrance, due to high percentage of citral in the oil. The characteristic fragrance of oil makes its use in scenting of soaps, detergents, insect repellent preparations *etc.* The major use of oil is used as a source of citral, which goes in perfumery, cosmetics, beverages *etc.* The citral rich oil has germicidal, medicinal and flavouring properties. During early fifties India had monopoly both in production and world trade. Considering the bio-diversity in *Cymbopogon* spp. found in India; some allied spp. have shown to contain high value of nerolidal and farnesol in the oil. Obviously, varieties with these high value aroma compounds should be developed. The oil of lemon grass has high percentage of terpenes (limonene and myrecene), beside menthyl heptenone, linalool, geranyl acetate, nerol and geraniol. Further, citral can be converted into high value compounds like cintronellal, geraniol, geranyl acid and geranyl nitride but the processes are governed by patents.

CKP 25, RRL 16, Jama Rosa, Praman, Sugandhi, Krishna *etc.* are the varieties of lemongrass. The crop is propagated through seed raised in nurseries, 2.5 kg/ha. It is also vegetatively propagated by splitting the clumps into slips. They are planted at a spacing of 40x40 cm. There are two planting seasons of lemongrass *i.e.* February-March and September-October. About 55,000 slips are required for one ha. A healthy plant gives about 100-200 g of seeds. These dry seed lots are stored in gunny bags lined with polythene. The seeds lose their viability if stored for a period more than one year. Lemongrass crop is free from most pests or disease but may require micronutrients over marginal lands. The field is kept weed free for the first 3 - 4 months after plating. Similarly, weeding-cum-hoeing is done up to 1 month, after every harvest. Generally, 2-3 weedings are necessary during a year. Under normal conditions, three harvests are possible during the first year and 3-4 in subsequent years, depending on the management practices followed. The yield of oil is less during the first year but it increases in the second year and reaches a maximum in the third year; after this, the yield declines. Under irrigated conditions from newly bred varieties an oil yield of 150-200 kg/ha is obtained. The oil is yellowish in colour having 75-85 per cent citral and small amount of other minor aroma compounds. The recovery of oil from the grass ranges from 0.5 - 0.8 per cent. If grown commercially, lemongrass can be a boon to the farmers of Jammu region. It can change the life of the farmers of area.

Chapter 17

Potential of Bamboo Cultivation in J&K

Bamboo is an important plant gifted by nature in tropical and subtropical areas of the world. It is also called as wonder plant, green gold, emperor of all grasses, wood of the poor *etc.* It is used by the people in their every-day utilities. It is closely interwoven with the life of the people in several ways. It can be used as a food, fodder, construction material, paper, mats, musical instruments, paintings *etc.* In India bamboo is expanding its footprints across the globe. India possesses one of the largest bamboo resources in the world. India has about 130 species covering 9.57 million ha area. It can be grown in between the other trees. In the event of a fire bamboo has the innate ability to regenerate and again grow green shoots next year. Bamboo has multiple economical, ecological and social values. It has the capacity to provide the ecological, economic and livelihood security among the people. It acts as source for the raw material for the industrial and domestic use. Bamboo biodiversity can be observed in different areas of J&K. Bamboo forms the most diverse group of plants in the grass family. It belongs to the sub-family Bambusoidae of the family Poaceae. It is a form of grass but it is strong and elastic. Bamboo is found in all the continents except Europe. It is probably the best natural engineering material on the earth. It mainly grows in the tropics and sub-tropics. It has versatile, enduring and highly renewable material that people have been utilizing for thousands of years. It can be harvested in 2-3 years. It protects the environment by reducing air and water pollution due to its excellent pollution absorption qualities. It produces more than 35 per cent of oxygen than hardwood trees. It can contribute in the economy and ecology of J&K. It can be used as the building material and also for scaffolding, building of bridges *etc.* Bamboo maintains its core temperature which helps to grow it in diverse climates. Bamboo occurs in different bioclimatically defined forest types ranging from tropical to subalpine zones. India is the second richest country in bamboo genetic resources after China. These two countries together have more than half the total bamboo resources in the world. About 136 species of bamboos are reported in India. Bamboo is grown on about 9.57 million hectares in India which constitutes about 12.8 per cent of the total area under forests. Out of the 22 genera available in India, 19 are indigenous and three exotic. The annual production of bamboo in India is about 4.6 million tonnes. The annual yield of bamboo per hectare varies between 0.2 and 0.4 tonnes depending upon the intensity of stocking and biotic interferences. The economic impact of the agro-forestry-based bamboo system may influence general economic development of the country. It is mostly found in N-E States, Chhattisgarh and Andaman and Nicobar Islands and Jammu and Kashmir. Bamboo is not infested with the insect-pests. After two years of unhindered growth, the bamboo grove can be harvested with the minimum labour. It is a very hardy grass which survived even after the nuclear blast at Hiroshima in Japan in 1945. It is a carbon-friendly grass and it can capture the carbon faster than any tree. Bamboo supports other agro-industries for manufacturing handicrafts items, packaging materials, raw materials for fencing, paper industries *etc.* across the globe.

Agriculture plays the crucial role in the economy of Union Territory of J&K. The agro-diversity of J&K varies from sub-tropical in Jammu region and temperate in Kashmir region. Agriculture is the dominant occupation in J&K and it provides employment directly or indirectly to about 70 per cent of its population. The average size of land holding of J&K is only 0.67 hectare against 1.33 hectares' land holding size on national basis. About 30 per cent of the area of J&K is under cultivation. The average annual rainfall of these two regions is 1069 mm and 660mm respectively. The average temperatures of these two regions are 24.5 and 13.3 Centigrade respectively. There is need to increase the income and employment in agriculture and allied sectors for the farmers of J&K. This is possible by searching some new avenues in agriculture. Union Territory of J&K lies in the North West of the country. About 19.95 per cent of the total geographical area of this hilly UT is covered under forests. The national level goal in forestry is to cover at least two thirds of the total area in the hilly regions under forests and trees cover. J&K has adopted the forest policy to maintain the adequate forests in various regions of the UT. There are very good reserves of bamboo in J&K and industries such as Agarbati can be established. Bamboo has steady demand over the years and its market value is not much affected.

There is dire need to mainstream the bamboo cultivation in J&K so that it can help to raise the socio-economic status of the farmers. There is also need to popularize bamboo cultivation specially in the wastelands as bamboo cultivation has immense potential in J&K. J&K already has some bamboo growing pockets where entrepreneurships can be generated by the value addition in bamboo. J&K has immense scope of agro-ecotourism, handicrafts, agarbati making *etc.* and bamboo can play an important role important role in these industries. Bamboo cultivation has number of advantages and its cultivation reduces the soil erosion. Bamboo cultivation can generate extra income and employment for the farmers. Bamboo has manifold uses from cradle to coffin. Bamboo has many uses as a substitute for fast depleting wood and as an alternative to more expensive materials. It can be used in paper industry, to make small footbridges, water pumps, irrigation pipes *etc.* In rural areas it can be used as fodder, in making brooms, food, buckets *etc.* Nutrient analysis of edible bamboos indicated that it contains 77 per cent to 84 per cent moisture, 4 per cent to 7 per cent protein, 5 per cent carbohydrates, 12-13 mg/100G vitamin C and Na, K, Ca, Mg and P as minerals. Bamboo has the potential for checking soil erosion, fast vegetative growth cover to deformed areas and road embankment stabilization. Bamboo tensile strength is 28000 per square inch versus 23000 for steel. It can be used for making pillars, roofs, ceiling, walls *etc.* It has been traditionally used to make medicines to treat asthma, cough and fever *etc.* It is also used to make certain structures in earthquake prone areas, in music and arts.

Bamboo is one of the fastest growing plants on the earth. Its growth rate ranges between 30cm to 100cm per day. It generates more oxygen than equivalent stand of the trees. It protects against ultra-violet rays and is an atmospheric and soil purifier. A story tells that once a forest officer kept his hat on a fast growing bamboo culm and forgot to take it with himself. The next day, he saw that the hat was beyond his hand's reach in 24 hours. Government is endevouring for bamboo cultivation. Planning Commission has set up a National Mission on Bamboo Technology and Trade Development. It has been realized that bamboo can help to uplift the socio-economic status of the farmers. The major objective of National Mission on Bamboo is to implement an action plan for keeping bamboo as a key component in the national scene to generate employment, to protect environment degradation and give a boost to bamboo industry. Alcoholic beverages can be made from bamboo leaves, charcoal can be made from bamboo, laminated bamboo mats can be made for disaster prone areas. Thus, bamboo can play a crucial role in strengthening the rural economy of J&K. There is need to form the bamboo clusters in J&K. Bamboo industries on different products needs to be established in J&K. Capacity building on value addition in bamboo can bring new opportunities of income and employment among the farmers of J&K. Bamboo cultivation in J&K can also be promoted through agro-ecotourism as some areas of J&K are visited by the tourists throughout the year. Bamboo growers need institutional support at regional, national and international levels. During the COVID-19 pandemic the information to the bamboo growers can be disseminated by the effective use of Information and Communication Technologies (ICTs).There is need to preserve the rare and threatened species of bamboo. A linkage needs to be developed between the bamboo growers and scientists. Private should also be involved in the bamboo cultivation. Appropriate species of bamboo are must for its promotion in J&K. More research is needed to specify the species of bamboo in particular belts of J&K and tissue culture labs can also be established. Bamboo technology parks can be set up in J&K. Bamboo model villages needs to be made for promotion of bamboo cultivation in J&K. The cultivation, buying, processing and transporting of bamboo needs to be streamlined in J&K. Proper marketing strategies need to be developed for the sale of bamboo products.

Chapter 18

Mushroom Cultivation as an Enterprise in J&K

In India, majority of population still dependent on agriculture and it is estimated that more than 600 million tonnes of agro-wastes are produced annually. The agriculture, which is an engine of growth and development and a significant contributor to national economy, has been greatly influenced by the process of globalisation. There is a matter great concern about imbalance of total production, the urban-rural divide, national food security and economic access to food. The agricultural strategy in the country seeks to bridge the product and production gaps. The policy envisages promotion of sustainable agriculture through a regionally differentiated approach, improvement in the input use efficiency, development and transfer of technology. There is a need to focus on technology generation and its application in agro-ecological or social circumstances.

Mushrooms are the fleshy-to-tough structures which are the part of the reproductive phase of a certain group of fungi. Mushroom cultivation was started in the 16th century but on a commercial scale it was started in Europe in 17th century and many farms for the mushroom production were established. India with the diverse climate conditions and abundant agricultural wastes has been producing the mushroom from the last four decades. Commercial mushroom cultivation in India has started recently and growing mushroom under controlled condition is of recent origin. Its popularity is increasing day by day and it has become a business which is export-oriented. Today mushroom cultivation has been taken up commercially in states like Uttar Pradesh, Haryana, Jammu and Kashmir, Rajasthan, *etc.* (during winter months) while earlier it was confined to Himachal Pradesh, J&K and other hilly areas. Mushroom is an excellent source of proteins, vitamins, carbohydrates, fibres, minerals, folic acid and is a good source of iron for anemic patients and is described as one of the best options to convert agro-wastes into quality proteins. Mushroom can be taken by the diabetic and heart patients as it has low calorific value. Mushrooms are of different types: (a) Button Mushroom (*Agricus bisporus*), (b) Dhingri (*Pleurortus*), (c) Milky mushroom (*Calocybe indica*). Of all the types, button mushroom is the most popular one (85 per cent). Mushroom cultivation can be done at cottage and small-scale levels besides large-scale farming. Mushroom cultivation provides option for income and employment generation specially for the educated youths and women without any major arable resources. China claims to grow 60 types of mushrooms and the reported production of all mushrooms put together was 18.2 million tonnes in 2008 which was around 70 per cent of world production.

Mushroom cultivation in the Union Territory of Jammu and Kashmir was started in 1964 by the Department of Agriculture, Lalmandi, Srinagar and work on edible fungi was taken by RRL at Srinagar. Presently around 600 tons of mushrooms is being produced annually in Jammu and Kashmir especially from Jammu, Kathua, Reasi and Udhampur. There is immense scope for round the year mushroom cultivation in Jammu region because of diverse agro-climatic conditions and availability of raw material. The farmers are showing keen interest in mushroom cultivation because mushroom cultivation generates more income as compared to other crops. The three types of mushrooms vary in their shape, size, colour and biochemical composition. Earlier only the button mushroom was grown but now oyster and milky mushrooms are also grown in J&K.

The main consumers of mushrooms are food restaurant, hotels, clubs and households. Mushrooms are mostly sold through vegetable shops. The growing domestic and export market as also the delicacy and food value provides extensive and good potential for cultivation of mushroom. Marketing is an Individual affair in our country and there are subsidies for establishing Canning Units and other such products but the marketing channels are limited. The seasonal growers do not have cooperatives. There is demand for opening the market outlets but this is feasible only when there is ample mushroom production. It has been observed that the mushroom growers lack basic knowledge of production technology and marketing channels. Mushrooms have perishable nature and this is the reason that the trade of mushrooms is limited. The canned mushrooms can be kept for a long time and can be transported in a better way. The USA and European Union are the major producers of mushroom but these are the major importers of mushroom as there is great demand of mushroom in these countries. China, India and Indonesia are three most important mushroom exporting countries in the world. Initially there were very few quality strains of mushroom. S-11 was introduced in mid 60s continues in use for 2-3 decades and is still used by the seasonal growers. Presently Krishi Vigyan Kendras (KVKs) of State Agricultural Universities (SAUs) and Horticultural Boards are playing a big role in popularizing mushroom cultivation among the farmers.

Spawn is mushroom seed. Spawns are available in the Directorate of Agriculture and State Agricultural Universities (SAUs). If desired, the same can be produced and

sold commercially. There are two methods of composting for mushroom cultivation. One is long method and another is short method. Short method requires less time but it is costly. Most of the growers still use the long method for the compost making which leads to the poor yield and also requires indiscriminate use of chemicals. There is need to diversify and popularize different types of mushrooms and go for multifunctional use of mushroom tunnels. These tunnels should not only be used for pasteurize compost for button but can also be used for pasteurizing straw for oyster, milky and button mushrooms cultivation. This will help in product diversification and there will be less use of power.

There are several mixtures for compost formation and anyone that suits the entrepreneur can be chosen. It is prepared by using wheat/paddy straw into which various ingredients are added. In synthetic compost wheat straw is supplemented with poultry manure, urea, gypsum and mustard oil cake. In organic compost, horse dung is added. The compost can be prepared by the long or short composting methods. Only those who have pasteurizing facility can employ short cut method. In long method 7-8 turns at regular intervals are required for a period of 28 days. Good compost is dark-brown, ammonia free, little greasiness and having 65-70 per cent moisture. Mushroom seed is said to be spawn which is white in colour. It should be free from any disease and insect-pests. Before spawning the surface should be washed with 2 per cent formalin. For mixing spawn with compost any of the three procedures can be followed: (i) Compost is divided into equal layers and spawns spread in each layer. Result is spawning in different layers. (ii) 3 to 5 cms of compost is remixed, spawns spread and covered with compost. (iii) Spawns are mixed with compost and pressed. A bottle of spawns is good enough for 35 kg of compost spread over 0.75 sq.mt. area (about 2 trays). That is, spawn to compost ratio is 0.5 per cent. Trays are then arranged in tiers in the cropping room and covered with newspapers. 2 per cent formalin is sprinkled over them. Desired room temperature should be around 18 degreeC with 95 per cent humidity.

Spawned compost is covered with suitable material is said to be casing. It helps to convert vegetative phase into reproductive phase. Besides temperature and humidity mentioned above, proper room ventilation should be ensured. After casing mushroom bags should be kept moistened. Mushrooms prop up in 30-35 days. These fungal fruit bodies appear in flushes and harvested when buttons are tightly closed. In a cropping cycle of 8-10 weeks an average yield of 10 kg mushroom/sq. metre is feasible. Cropped mushrooms can be packed for marketing.

Technical training and assistance can be taken from. (a) Krishi Vigyan Kendras (KVKs) of State Agricultural Universities (SAUs) (b) Rashtrya Anusandhan Evam Prashikshan Kendra, Chambaghat, Solan (H.P.). (c) Regional Research Lab, Jorhat, Assam. (d) Regional Research Lab, Jammu. (e) Central Food Technology Research Institute, Cheluvamba Mansion, Mysore *etc.* Raw materials should be procured preferably from local areas. It has been observed that the mushrooms are having good demand in the market. Mushrooms cultivation requires awareness on modern production technology and health benefits, product diversification, better transportation with cold chain facilities, cooperatives, incorporation of mushroom in Indian food habits, introduction of mushrooms in mid-day meal scheme in schools and dissemination of information for scientific mushrooms cultivation through electronic media. There is ample scope of integrating mushroom farming in existing integrated farming system. There is need of simple scientific technology, better strains and more value addition methods of mushrooms for the mushroom growers. There is need to organize the trade and marketing of mushrooms within J&K and country.

Chapter 19

Combating Monkeys Menace for Rural Prosperity in J&K

Monkeys are wreaking havoc in the Union Territory of Jammu and Kashmir. Their burgeoning population and habitat shift from forests to towns has created a new level of man-animal conflict in this hilly UT. Monkeys damage the crops and leave farmers grappling with losses that have badly affected J&K economy thus affecting the livelihood of the farmers. Due to the damage of agricultural crops some farmers are abandoning agricultural occupation and searching some other avenues. Some farmers say that it is better not to cultivate their fields because it saves their input costs. Our neighboring state Himachal Pradesh used to sterilize the monkeys to check their offspring. It has been observed that the destruction of crops and geographical reach of the monkeys is increasing annually. Already J&K is having low crop productivity due to natural calamities, rainfed conditions and low adoption of advance agricultural technologies by the farmers. Crops damage by the monkeys is a matter of grave concern and is discouraging the farmers to grow the crops. Even if the high yielding/hybrid varieties are sown by the farmers, monkeys finish the crops beforehand. Monkey menace has become a big issue in the rural areas of J&K.

Some areas of J&K are severely affected by the monkeys due to which farmers are not sowing crops in their land resulting huge economic loss to them. Monkeys invade the fields and severely damage the maize crop, vegetables, fruits *etc.* In Jammu region of J&K every year crops are severely damaged in around 15,000 hectares of land in around 250 villages of Reasi, Udhampur, Jammu and Kathua districts of J&K by the monkeys resulting in the loss of crores of rupees. In Kashmir region of J&K there is less threat of the monkeys. Monkeys have left a substantial dent in J&K agricultural production capacity. Monkeys also cause substantial loss to the fruit crops like mango, guava, bear, grapes, citrus fruits, litchi, pear, peach, plum, apricot *etc.* in large area of J&K resulting a great loss to the orchardists. It has been observed that areas near the forests are less affected by the monkey menace as the monkeys get food in the forests. The crops like turmeric, ginger, marigold, lemon grass, palma rosa *etc.* are not affected by the monkeys and needs to be promoted in J&K. Fruit loving trees which are preferred by the monkeys needs to be planted along with periphery of forest and protected areas boundaries to augment the food availability for them so that they remain restricted in the forest areas only and do not come out in the villages to damage the agricultural crops.

The numbers of reported monkey attacks on humans are also rising. Sometimes some people die due to their attack while some are disabled after being attacked by them. They bite people, contaminate drinking water in storage tanks, transmit diseases, steal, tear, break, snatch and enter houses, buildings and even hospital wards. Our neighbouring state Himachal Pradesh has established eight sterilization centres to tackle the monkey menace. After the sterilization, the monkeys are released in the forests. The farmers in Himachal Pradesh are also using electric fencing that is absolutely safe and does not kill the monkeys but just scare them. After mild current the monkeys afraid to come in the fields. Forest department of Himachal Pradesh is using GPS to track the monkeys by plotting them directly on a map. They are also planning to install microchips in the sterilized monkeys for better identification.

Painthal, Dera Baba and Pouni areas of Reasi district of J&K are severely affected by the monkey menace. According to the locals some farmers are leaving the agriculture because of monkeys' threat. According to Sh. Ashwani Kumar, a farmer from Dera Baba village in Reasi district "We remain at maize fields right from morning to evening to protect our maize crop from monkeys which is a major food crop for us and we avoid to go even in our relatives home because of the monkeys. Our crops are destroyed by the monkeys and socio-economic status has totally been disturbed by the monkeys". According to Tilak Raj, a progressive farmer from Sirah village near Katra, there is dire need to protect our crops from monkey menace and some initiatives are urgently needed to manage this serious problem. In some areas of the state entire crops are damaged by the monkeys threatening the food insecurity to the human beings.

Meanwhile, the country's foremost wildlife research institute, the Wildlife Institute of India (WII), is working on oral contraceptives for monkeys that can be given through food. Experts say this can make capturing of monkeys for sterilization redundant. The major advantage of oral contraception is that it is non-surgical. According to the sources, Himachal Pradesh forest department has asked several agriculture universities including Chaudhary Sarwan Kumar Agriculture University, Palampur, to develop feed for monkeys in which these contraceptives can be given.

The HP government has also tried employing ultrasonic guns that can scare monkeys. Wildlife department officials say Van Vatikas are unsustainable because of infighting within various troops of monkeys. Himachal Pradesh

spending crores of Rupees to manage monkey population in the state. In order to relocating monkeys out of the city, the Delhi government hired Gray Langurs, the most common long-tailed monkey across south Asia, to chase monkeys. But the practice was discontinued after the Union environment ministry said it violated the Wildlife Protection Act, 1972, that says Langurs are the endangered species. The administration then tried to use young men dressed as Langurs to chase monkeys which had limited success. Telengana, which is home to more than 200,000 monkeys, has proposed to plant trees across the state and create green islands to provide habitation to monkeys. Government has announced the Rs 830-crore Haritha Haaram (Green garland) scheme. Under it, the state plans to its increase tree cover from the present 24 per cent to 33 per cent. The monkey's habitat and food-sources have been greatly depleted. Monkey society is built on a strong hierarchy of dominance, where the highest ranking is free to exploit all subordinates, and so on down the line of the hierarchy.

There is need to try to emulate the Himachal Pradesh's sterilization drive in J&K also and for that purpose special budget needs to be sanctioned by the state Government. The J&K government should also issue a notice that levies a fine on people feeding monkeys at roads, pilgrims and tourists sites. Feeding to the monkeys leads to abandon their natural habitat and they migrate towards villages and towns and create lot of problems to the people. Feeding, baiting and luring of monkeys are the offences under the Wildlife Protection Act. According to the experts the act of feeding makes them believe that humans are inferior to them. Combinations of strategies are needed to control the monkey menace in the state. Krishi Vigyan Kendra (KVK), Reasi is promoting the ginger, turmeric, marigold, lemon grass, vetiver grass and geranium grasses in the district which are not affected by the monkeys and farmers can generate extra income and employment by commercially growing these crops. Also there is need to restore the forests so that monkeys remain confined in those areas.

Chapter 20

Rural Tourism in J&K

Union Territory of Jammu and Kashmir is endowed with ample natural resources including soil, water, climatic condition, diversity, topography, rich natural flora *etc.* which are conducive for the tourism. Jammu and Kashmir is located between 32.17 and 37.06 North latitude and 73.2 and 80.36 East longitude in the Himalayan region. The agro-diversity of J&K varies from sub-tropical in Jammu region to temperate in Kashmir region. The average annual rainfall of these two regions is 1069mm and 660mm respectively. The average temperatures of these two regions are 24.5 and 13.3 Centigrade respectively. High Mountains and small hills, snow peaks, rivers and rivulets, glaciers, valleys and lakes, large forest areas and terraced fields are the integral features of hilly areas of Jammu and Kashmir. Jammu and Kashmir lies in the North West of the country. The industrial sector in J&K is not well developed and rural tourism can play a key role for the upliftment of the rural economy. Rural tourism can improve the quality of life of the rural people of J&K. Rural tourism entrepreneurship has gained importance as it is seen as a major driving force behind rural tourism. J&K has lot to offer to the tourists across the globe because of its scenic beauty, a kaleidoscope of traditions, a variety of cultures and an array of opportunities to explore the outdoors activities through sporting and adventure activities.

Jammu and Kashmir is one of the most beautiful tourists' destinations in the world. This UT is characterized by scenic beauty, folded mountains, deep gorges, natural waterfalls, alpine pastures, lush green forests, rich wildlife, snow clad fields, carpet green turfs, gushing fountains, lofty skying grounds, charming gardens, cool pollution free breeze, shimmering lakes, apple orchards, benevolent people *etc.* Tourism is an important economic activity in J&K. Around 20 per cent of the workforce of J&K is directly or indirectly dependent on tourism. Jammu and Kashmir is blessed with enormous sites which attract the tourists across the globe. Still there is much scope to accelerate the process of tourism in J&K. The rural areas need to be selected as new sites of attraction for the tourists so that the rural people are benefitted from the process of tourism. In order to bring the rural people on the scene of tourism, rural tourism is one of the important options for J&K and this can strengthen the rural economy. Under rural tourism villages are designed in such a way that fresh and non-polluting environment, traditions, village handicraft, village folk dance and music, painting and sculpture, history of the region, nature of local work and technology use, architecture, education system, religion, dress, leisure activities *etc.* can attract tourists from outside.

Rural tourism includes a range of activities, services and amenities provided by the rural people to attract tourists to their area in order to generate extra income and employment. The growth of new employment opportunities is lesser than population growth. Tourism forms an important part in J&K economy. A need has been felt to extend the tourism development in rural areas and to tap the natural rural beauty of J&K in the form of nature based projects. Industrialization and development is mostly urban centric. The urban stress has led to growing interest in the rural areas. Rural tourism is a new concept which covers all those aspects of travel whereby people from the urban areas learn the ways of life and thoughts of rural people. Rural tourism refers the tourists' interest in the customs of the indigenous and exotic people. This tourism showcases the rural life, culture; heritage and art of rural areas and rural people are socially and economically benefitted and enable the interaction between the villagers and tourists. It can also be termed as agricultural tourism, cultural tourism, nature tourism, eco-tourism *etc.* Rural tourism has certain characteristics like it exits in natural environment, sparsely populated, experience oriented and based on the preservation of culture. In this tourism, tourists visit the homes of local people to understand their customs, traditions, life style, food habits, their festivals, folk dances, songs, rituals *etc.* It also involves a study, research and purchase of local products. Rural tourism fascinates the domestic and outside tourists as villages in Jammu and Kashmir have their unique styles and splendid natural beauty. The charming beauty of the villages could be the centers of attraction for the people of urban areas who live in the congested and polluted cities. The urban people lifestyle is stressful due to their hectic lifestyle. They want to get away from their daily routine work and want to enjoy the natural beauties of rural areas during weekends and holidays.

Rural tourism can create the job opportunities for the rural people. Environmental consciousness, increasing interest in culture and heritage, improving villages' accessibility *etc.* are the factors responsible which are shifting the trend towards rural tourism. Under this scheme key geographic regions are identified for the development of rural tourism. Jammu and Kashmir has a great potential for rural tourism, due to its unique heritage, culture and natural attractions. Rural tourism is useful for J&K where around 80 per cent people reside in the villages. Rural tourism is participatory and designed in such a way so that the economic and social well-being of local people can be improved. It also points the need for careful planning to project the integrity of sites and highlight the village culture. It is planned in such a way that local community needs to ensure that rural tourism sustains and benefits local community socio-culturally and economically and is not simply the gaining the money for government or private entrepreneurs. If we want to commercialize the rural tourism, we need to endeavor to satisfy its obligations towards local villagers. Rural tourism is a socio-cultural event for tourists. Through rural tourism one gets the opportunity to enjoy the natural beauty of villages, observes different cultures and way of life of the rural people.

Tourists are a major factor of socio-cultural change. Rural tourism has many advantages. It enhances communication skills, improves self-image of the community, increases employment opportunities, improves educational opportunities, motivates for learning new things *etc.* of the rural people. Rural tourism creates new jobs in rural areas, attracts many youths for rural employment and slows down the migration of rural people in the urban areas. New jobs in the rural areas may motivate the rural people for education. Employment opportunities in rural tourism can divert workers from other professions. Employment generated in the villages may improve the status of the rural people and can also increase the income of rural people. Rural tourism can create confidence among the rural people and revive the local people's interest. Rural people may have pride in their culture, traditions and values. Mass media and hegemonic culture can promote the local folk traditions into oblivion. Rural tourism can also provide the opportunities to study the religious and ritualistic services of the rural areas. These aspects of a culture can fascinate the tourists. Religious rituals, festivals and public performances can provide inputs to symbolic qualities of rural tourism.

Rural tourism is an invisible export. It is different from other sectors of industry. The rural tourism industry can earn a huge amount of money. Rural tourism is thus a source of aid from within J&K, country and out of country. Rural tourism is directly helpful in correcting UT's economic position and can divert the rural people towards progress. Through rural tourism government can gain more economic profits from taxes on things tourists use and buy. Rural tourism has now become a major sector influencing all aspects of a nation's. Government should take proper measures for creating framework for the promotion of rural tourism. There is need to increase the rural tourism at a fast rate and it should be diversified. Increased tourists traffic in J&K calls for new marketing and pricing policies. There is need of proper planning and effective legislation for resource conservation and balanced developmental activities for the rural tourism. The technical resources are also required for successful promotion of rural tourism. Rural tourism can be fostered if it is an integral part of a balanced economic and social development programme. It is not only the government but the private sector should also play an equal responsible role for the development of rural tourism. Rural tourism can make our rural areas become attractive and prosperous.

Chapter 21

Parthenium: A Major Threat to Agrobiodiversity of J&K

Union Territory of Jammu and Kashmir is situated in the Northern most part of India and its major part is situated in the Himalayan region. This UT is blessed with immense natural beauty and is considered as the heaven on the Earth. Most of its land is under orchards, pastures, grasslands, forests and wasteland ecosystems. As most of these lands are not used for frequent cultivation, the obnoxious weed like *Parthenium hysterophorus* (Congress grass) has invaded most of these lands. Parthenium weed is regarded as the worst weed because of its invasiveness, potential for spread and economic and environmental hazards. This invasive weed is presently almost found in all the parts of J&K causing a serious threat to the biodiversity of this beautiful part on the Earth. This weed is considered as the dominant weed in J&K as it has covered almost all the fellow lands, river beds wastelands, pasture lands, grasslands *etc.* It is a noxious weed because it is highly adaptable to almost all types of environmental conditions and can invade all types of lands, causes high losses in the yield of crops. This weed forms dense, impenetrable thickets and reduces the productivity of crops, pastures, orchards and forestry plantations by its competition for resources and allelopathic effects. The low crops production and productivity due to this weed leads to scarcity of food, fuel wood, fodder, fruits, increase in monkey menace and migration of rural people to urban areas in search of employment after leaving the land fallow.

About 80 per cent of population of J&K directly or indirectly depends on agriculture. People of J&K fulfill their subsistence needs from cultivated, uncultivated and degraded lands. Crops production and productivity can be enhanced in J&K by the management of this problematic weed with the effective technologies. There is need to understand the biology of this obnoxious weed, its ecological impacts and management techniques. This troublesome weed has high rate of dispersal and adaptation to adverse conditions. About 30 per cent of the land of J&K is under cultivation. This weed is now spreading rapidly its tentacles in agricultural lands, forests and pastures. It is spreading at an alarming rate in the J&K and is found in almost all the districts of J&K. This weed is inversely affecting the biodiversity and ecological system of the J&K. It is said that it was introduced in Jammu and Kashmir in 1963 from Madhopur in Punjab across the Ravi River all along the national highway. After that it has been dispersed in all over the state. Presently its infestation is alarming as it has covered large area in J&K. It has been observed that this weed has reduced different crops yields and has also affected biodiversity in J&K. It can be seen on roadside, railway tracts, vacant lands, wastelands, agricultural, horticultural and plantation crops, industrial areas, irrigation canals *etc.* in almost every district of J&K.

Parthenium hysterophorus (Congress grass) is commonly known as carrot weed as its plant appears like carrot plant. It is herbaceous, an annual plant belonging to subfamily Heliantheae and family Asteraceae (Compositae). Its vernacular names are as gajar ghas, ragweed, white cap or top, gajari, chatak candani, nakshtra gida, safed topi *etc.* This weed has been rapidly spreading from the last two decades across the J&K and now this weed has attained the status of "Worst Weed" because of its allelopathic effects on different crops and harmful effects on human beings and animals. It has been observed that its infestation causes crops yield losses up to 40 per cent in several crops and reduces fodder production up to 90 per cent. It is an aggressive annual herbaceous plant which has been widely dispersed across J&K. This weed rapidly covers the new surroundings and poses a serious threat to the environment and biodiversity of J&K.

This dangerous weed has been categorised as "Cosmopolitan weed", "National culprit" and "National health hazard" due to its serious environmental threats. Parthenium origin is considered to be Mexico. In India, its occurrence was first noticed in Pune (Maharashtra) in 1955 and now it has covered almost all parts of the country. It is supposed to be introduced in India from the United States of America along with wheat and other cereals import. By 1972, it had dispersed into the majority of the Western states/UTs from Kashmir in the North to Kerala in the South. It has widely spread in India from Kargil region of Jammu and Kashmir to Port Blair in Andeman and Nicobar. This weed grows at a faster rate due to its aggressiveness, high vitality of seeds, easily dispersal, innate dormancy and prolific seed producing nature. Any part of the plant (even root) can cause the subsequent risk of allergic reactions. In humans it causes health hazards like skin allergy (dermatitis), hay fever, asthma and bronchitis with flowers, seeds and even hair on leaves. The skin becomes photosensitive making it hard for the patient to face the sunlight. Bare parts of the body, soft tissues around the eyes and under joints are more likely to be effected by the contact dermatitis. Allergic papules are observed in school boys when they volunteered for uprooting parthenium. Animals are equally prone to the harmful effect of the weed. During summer seasons, when the animals do not

get palatable species in grazing lands, they are forced to feed on parthenium. As a result milk taste becomes bitter and they suffer with ulcers in mouth and intestine. Whenever animals walk or graze through parthenium, their udders are inflamed and they suffer with fever and rashes. Histopathology of the kidney and liver revealed degenerative changes and necrosis. Being toxic to livestock,causing both acute and chronic toxicity, a noticeable reduction in milk yield, tainting of milk with parthenin, depigmentation of skin, tainting in mutton and bitter taste of milk have been observed. The milk consumption of the animals grazing around parthenium invaded fields is hazardous to man. Some animals feeding on parthenium die due to acute dysentery, itching, cryhematous, development of oedema around eyelids, dorsum of tongue, loss of hair *etc.* Parthenin is the chief chemical (0.3 per cent) found in the weed.

Community efforts involving all sections of the society are needed to manage the parthenium. Uprooting the weed manually when the soil is wet and slashing with word, collecting and burning the weed before flowering are some of the means of its manual control. Spraying of a solution of common salt (Sodium Chloride) at 15-20 per cent concentration has been found very effective. Applications of herbicides like glyphosate (1-1.5 per cent) for total vegetation control or metribuzin (0.3-0.5 per cent) if grasses are to be saved in non-agricultural land are considered effective in preventing this weed spread. It can also be controlled by the use of bio agent Mexican beetle (*Zygogramma bicolorata*) as it is natural, self sustaining, inexpensive and is ideally suited to non-crop situations and wastelands. This bio agent remains most active during rainy season and it completely controls the weed. The plant species like Cassia tara, Cassia seriea, Amaranthus asper, Malva pustulata *etc.* have capability to replace parthenium. The other way to manage parthenium is by uprooting it before flowering and make vermicompost by pit method. Vermicompost can be applied in different crops to mitigate the nutrients problems. It can also be used in papermaking, an antifeedant and phagostimulants. National Research Centre for Weed Science, Jabalpur (M.P.) organises many awareness programmes on Parthenium eradication throughout the country by involving Krishi Vigyan Kendras, institutes under ICAR, environmental agencies, NGOs *etc.* Various posters, folders, technical bulletins, books. Parthenium eradication has become a challenge in J&K because of its epidemic proliferation and strong reproductive potential. This weed is spreading at an alarming rate in J&K and some strenuous efforts and strategies are needed to eradicate this menace so as to improve the production and productivity of different crops. Both public and private organizations need to work together to manage this obnoxious weed.

Chapter 22

Protection of Green Gold in J&K

Union Territory of Jammu and Kashmir lies in the North West of the country. About 19.95 per cent of the total geographical area of J&K is covered under forests. Around 59 per cent of the geographical area of J&K is under permanent snow cover, glaciers and cold desert and thus unable to support trees growth. The national level goal in forestry is to cover at least two thirds of the total area in the hilly regions under forests and trees cover. Jammu and Kashmir regions forests account for 47 per cent of the geographical area. Jammu and Kashmir adopted its own forest policy to maintain adequate forests in various regions of J&K in 2011.Jammu and Kashmir is one of the most beautiful tourists' destinations in the world. J&K is characterized by scenic beauty, folded mountains, deep gorges, natural waterfalls, alpine pastures, lush green forests, rich wildlife, snow clad fields, carpet green turfs, gushing fountains, lofty skying grounds, charming gardens, cool pollution free breeze, shimmering lakes, apple orchards, benevolent people *etc.* Forests are useful for human livelihood, soil and water conservation, biodiversity, green house gases, flood control *etc.* Forests were generally vested with the government especially after 1947 but usufruct rights of local communities, ethnic groups and tribes continued to be part of management. Irrespective of ownership, forests were treated as open or community resources of local population in the limited sense of traditional rights.

Nature has bestowed Union Territory of J&K with a varied geographical and ecological richness, making it a great reservoir of biological paradise on earth. Jammu and Kashmir is a unique forestry UT which is having sub-tropical and temperate regions. Around 36.50 per cent of the flora of the country is endemic to this region, which is of great significance. Forests affect nearly every aspect of our life and are very important for the environmental, ecological and economic balance of our society. Forests are rich in biodiversity and are also the important sources for creation of employment and livelihood for large number of people of J&K. Forests are also prime source of fresh water for both urban and rural population as they are helpful for the rainfall. It has been observed that the density of the forests is decreasing at an alarming rate and many important species of the plants/trees have extinct and many more species are in endanger. Now-a-days we observe that the illicit trade of forest wood and encroachment of the forest land by the people by which the forests are shrinking day by day. The flora of the prime species of the forests is reducing. In order to safeguard them from biotic interferences strenuous efforts are needed by both the government and non-government agencies. Increasing population, lack of awareness of the people pertaining to forests, hunger for land grabbing, unsystematic developmental activities have taken back lot of valuable genetic resources by the forests. Ecological studies regarding the continuous growth, stability, evolution of the ecosystem and organisms are necessary for healthy environmental management and also to balance the finite resources present on the earth. The available resources should be protected from degradation. Man himself has adversary affected the benefits provided by the forests to the living organisms. Following strategies are needed for the conservation of forests for the safe future:

Afforestation

Endeavors must be made to increase the area under forests and tree cover through implementation of different afforestation programmes/schemes.

According to the National Forest Policy (1998), 66 per cent of the geographical area of the hill states should be brought under forests but the presently our state is well short of this (only around 19 per cent). Roadsides plantation is must to reduce the level of the noise and the air pollution.

Biodiversity

Biodiversity and wildlife conservation and management is supported by central legislations such as biodiversity Act, 2002 and Wildlife (Protection) Act, 1972 and Statutory bodies such as the Biodiversity Action Plan by the J&K Govt. through, J&K SFRI vide Govt. The J&K government must be committed to conserve and manage the threatened biodiversity of the state. The management of the existing protected area network must be strengthened to conserve and increase the genetic stock of the species that are specifically endangered. The Biodiversity Action Plan should be properly implemented. Measures for the ever-increasing dynamics of the human- animal conflict should be contemplated on priority basis, as this is the most challenging aspect in present era.

Medicinal and Aromatic Plants

J&K is said to be treasure house of many medicinal and aromatic herbs as the hilly areas are the hubs of these plants. There is dire need of conservation of these medicinal and aromatic plants in J&K. Close co-ordination is needed with the state agricultural universities and other forest research institute for the intensive research on this issue. Indigenous Technical Knowledge (ITK) practices of medicines must be encouraged and should be properly documented.

Forest Fire Management

A forest fire causes irreparable damage to wildlife, biodiversity, water resources, forest based livelihoods and well being and changes heaven into hell within very less time. Fire control measures such as creation of fire lines, effective information and communication technologies (ICTs), defined responsibilities and introducing the incentives to the local communities living in the fringes of these forests for participation in the fire management must be developed. Training programmes with regard to fire fighting to government employees, local communities and interested rural youths/persons in order to create more awareness on causes and the control of forest fires must be arranged

Watershed Management

It is an integral part for the sustainable forest management. The management of natural resources including the water management is of paramount importance in present era forestry system. It will help to check soil erosion and denudation of the catchment areas of the river systems of the state for mitigating the floods, landslides and siltation. Field programmes must be launched to promote protection and the development of the water catchments of J&K. Effective measures must be taken for promoting harvesting and conservation of the rain water.

Eco-tourism

Eco-tourism has immense scope in J&K as it will help to increase income and employment in the rural areas. Forest areas having the potential of eco-tourism must be established by developing eco-tourism cell in J&K. While developing eco-tourism, emphasis must be given on ecological and environmental concerns of J&K.

Research and Development

Research improves and develops new knowledge about the effects of different activities. The main thrust should be given on forestry research in J&K. Dynamic scientific human resource persons must be recruited who will assist to the process of developing short and long term agenda with special focus on applied forestry research. There must be collaborative research between state forest research institute and Forestry Divisions of state agricultural universities for effective benefits. Extension publications regarding the knowledge of forests should be published for creating the awareness among the general public. The research journals must also be published from the forest research institute and also from the state agricultural universities so that the transfer of research findings and technological skills/know-how must be disseminated to the people. Modern nursery technologies supported by the sound silvicultural practices must be developed to prepare the quality planting material. J&K government is making efforts for the conservation of forests but more efforts are needed for shaping the forest policy of J&K which may help to encourage the protection, conservation and enrichment of the forests.

Chapter 23

Shafanoo: Emerging Bamboo Village in J&K

Union Territory of Jammu and Kashmir lies in the North West of the country. Bamboo cultivation has immense potential in J&K. There is need to popularize bamboo cultivation particularly in the wastelands of J&K. It can be a source of extra income and employment for the farmers of the J&K. Bamboo production is a highly commercial industry and its cultivation reduces the soil erosion. Reasi is a hilly district of Jammu and Kashmir which is located around 80 km away from Jammu and is at 1528 meter above mean sea level. It was carved from Udhampur district on 1st April, 2007.This district is surrounded by Udhampur, Rajouri, Jammu, Ramban and Shopian districts. This district of Jammu and Kashmir is world famous because it is an abode of Shri Mata Vaishno Devi Shrine. The Chenab River passes through this district which is another centre of attraction in the district. Reasi district is having immense scope in bamboo cultivation and there are certain areas in Reasi district where bamboo is grown commercially. Most of the farmers of Reasi district are small and marginal and crops are mostly cultivated under rain fed (about 94 per cent) conditions. Major crops grown in this area are maize, wheat, paddy, mash and potato. Farmers of this district are having very limited resources and mostly they adopt the subsistence system of agriculture. The yield of all the crops is low as compared to the national and state average yields. The adoption percentage of modern agricultural technologies is low resulting in the low socio-economic status of the farmers. There are various insect-pests and disease problems in the district. Boars and monkeys too damage the crops to a large extent in this particular district causing a huge loss to the farmers. Agroforestry has immense potential in the district as large part of the district comes under forests. If grown scientifically, bamboo can play the pivotal role in enhancing the income and employment among the farmers of the district.

Bamboo is an important plant gifted by nature in tropical and subtropical areas. It is also called as wonder plant, green gold, emperor of all grasses, wood of the poor *etc.* It is used by the people in their every-day utilities. It is closely interwoven with the life of the people in several ways. It can be used as a food, fodder, construction material, paper, mats, musical instruments *etc.* Bamboo has multiple economical, ecological and social values. Bamboo has manifold uses from cradle to coffin. Bamboo has many uses as a substitute for fast depleting wood and as an alternative to more expensive materials. It can be used in paper industry, to make small footbridges, water pumps, irrigation pipes *etc.* In rural areas it can be used as fodder, in making brooms, food, buckets *etc.* Nutrient analysis of edible bamboos indicated that it contains 77 per cent to 84 per cent moisture, 4 per cent to 7 per cent protein, 5 per cent carbohydrates, 12-13 mg/100G vitamin C and Na, K, Ca, Mg and P as minerals. Bamboo has the potential for checking soil erosion, fast vegetative growth cover to deformed areas and road embankment stabilization. Bamboo tensile strength is 28000 per square inch versus 23000 for steel. It can be used for making pillars, roofs, ceiling, walls *etc.* It can be used to make medicines to treat asthma, cough and fever *etc.* It is also used to make certain structures in earthquake prone areas, in music and arts.

Shafanoo village is about 40 km away from world famous religious place Katra in Reasi district of Jammu and Kashmir. The village is around 10 km. away from Krishi Vigyan Kendra (KVK), Reasi and farmers of this village seek technical guidelines from KVK scientists. This village is situated in the remote area of Reasi district and is lacking the road and communication connectivity. There are about 350 families in the village. The major occupation of the people is agriculture and the village is known for the commercial cultivation of bamboo. Most of the farmers of the village do the commercial cultivation of bamboo. There are three Govt. primary schools, two Angadwadi centres and one Janj Ghar in the village. The village is having around 250 hectares of land. There is no source of irrigation in the village and the village is 100 per cent rain fed. There are about 250 progressive bamboo growers in the village who are producing the bamboo at a commercial level. They have made furnaces in the village where the bamboos are moulded according to their use. The bamboos of this village are very strong and are used for the building purpose, baskets, stairs, nets, for flags *etc.* The farmers' sale their produce in local market, Jammu and even in Kashmir province at the rate of Rs.100-120/bamboo. Generally the farmers of this village fetch more money by selling their bamboos but this year due to COVID-19 pandemic their market has been influenced. Farmers also grow maize, wheat and few local varieties of vegetables as their forefathers were doing. Monkeys and wild boars are the great threats in their fields. Most of the agricultural produce is consumed by their families with very little surplus to sell. So bamboo is the main livelihood source of the farmers of this village. The technical guidelines on scientific bamboo cultivation are provided by the KVK, Reasi. With the generation of extra income, the farmers developed confidence to mitigate their basic problems. A self- reliance and entrepreneurial spirit has been developed among the farmers. Shafanoo farmers' success in commercial bamboo

cultivation is inspiring the other farmers of the area and they are also shifting towards the commercial bamboo for better economic returns. Gradually many farmers of the area are becoming interested in commercial bamboo cultivation. Bamboo cultivation helped to overcome the problem of monkeys menace in the village as bamboo crop is not damaged by the monkeys. According to a local farmer, Sh. Payar Hussain our village is known for bamboo production and quality bamboo is being produced by our village farmers. The village farmers want some bamboo processing units in the village so that income and employment can be generated in the village. It can generate more income and employment in the village. Farmers are showing keen interest on the capacity building for bamboo processing. According to the Naib Sarpanch of the village, Sh.Ratan Lal, "We face problems to sale our bamboo produce but if some industries on bamboos processing are made in the village then whole scenario of the village can be changed. New avenues for more income and employment can be generated for the villagers."

Many farmers grow bamboos in and around their farmlands to increase their farm income. The development of artisan skills for handicrafts and wider utility of bamboo can provide more employment opportunities and better income distribution for the people of this village. There is a dire need to aware the bamboo growers about the latest information on introduction, phenology, reproductive biology, propagation, exchange of germplasm, conservation status *etc.* There is also a need for the introduction of other species of bamboo of ethno biological significance. It has been observed that most of the bamboo growers are lacking up-to-date information on ecological and growth parameters, morphological variability's, conservation value, methods for multiplication *etc.* GIS based information system can be developed for the protected areas. Bamboo processing units can raise the socioeconomic status of the farmers of this village. Ethno biological approach can be adopted towards the cultivation, preservation of rare and threatened species of bamboo in this village. There is need of systematic research studies, range of distribution, flowering periodicity, ethno biological utilization, phenology, floristic compositions, introduction in production forests and homelands of rural areas. Inter-institutional support both at regional and inter-regional levels and interaction may help in developing, improving and managing the bamboo cultivation in this village.

Chapter 24

Turmeric: A Viable Crop in Monkey Menace Areas of J&K

Farmers of hilly areas of Union Territory of J&K face serious threats from monkeys as they spoil their many crops and thus directly affecting their income. More pronounced damage caused by these animals in lands adjacent to forest areas, due to food and water shortage in the forests. In Painthal block of Reasi district of J&K, it has been observed that some farmers have kept their lands fallow because of monkeys menace. They responded that their major crops such as maize, vegetables, fruits *etc.* are badly spoiled by the monkeys resulting a net loss to them. Some of the crops such as turmeric, ginger, marigold, lemongrass *etc.* were suggested by the scientists which are not affected by the monkeys. All these crops have immense potential in the region and can be grown by the farmers at a commercial scale. Turmeric of this area is famous as it has some typical characteristics regarding to its colour, fragrance, cumin content *etc.* If turmeric is grown commercially by the farmers of the area, they can fetch more money. Their fallow lands can be covered by this crop. Turmeric has tremendous potential in Reasi district as the tourists across the nation visit Katra throughout the area and there are numerous hotels and restaurants.

Scientific name of turmeric is *Curcuma longa* belonging to Zingiberaceae family and is considered as the triploid. In Hindi it is commonly called as Haldi. It was originated from the South Asia region. Its somatic chromosome number is 63. Turmeric also said to be the golden spice is one of the most important spices across the globe. For countless centuries, many different people are using this versatile herb to treat a myriad of ailments. This crop is known for its multipurpose value such as for the medicines, colour pigment, spicy flavor *etc.* It is anti-cancerous, anti-inflammatory, antiseptic, antispasmodic, antiproliferative, antioxidant, carminative diuretic *etc.* The curcumin found in it is used as a food colourant. Its annual growth rate in terms of area is 3.7 per cent and in terms of production is 9.1 per cent. Modern science has recognized the healing qualities of turmeric and much research is being conducted on it. Presently turmeric is being used in the treatment of the most intense ailments afflicting today such as Diabetes, Sclerosis, Sexually Transmitted Diseases, Irritable Bowel Syndrome (IBS), Acne, Kidney Infections, Alzheimer's Disease, Arthritis, Anemia, Leprosy *etc.* It can also be used as the mosquito repellent, cure of scorpion stings and wound healer. It also helps to balance the reproductive system of the females and males. Presently it is one of the most important herbs in any natural medicines. Turmeric was very sacred to the Aryans due to its golden yellow colour. Even now the Hindus consider turmeric to have auspicious qualities and use it in many sacred ceremonies. During the Indian wedding ceremonies, the bride and groom paste it on their bodies.

India is the largest producer, consumer and exporter of turmeric in the world and 46 per cent of world turmeric trade is run by India. Jamaica and Peru are the main exporters of turmeric while Iran is the largest importer. In India total area under turmeric is about 1, 72,000 ha and total production is about 8, 51,000 tonnes. Turmeric occupies 6.6 per cent of total area of spices in India. In order to mitigate the increasing demand of this spice there is dire need to cultivate this crop in a scientific way. Turmeric became valuable to human beings when it was discovered that the powdered rhizome preserved the freshness and nutritive value of foods. Turmeric is used as a condiment, dye, food colourant, drug and medicine. Turmeric rhizomes have yellow colour component as curcumin (3-9 per cent), essential oil (5-9 per cent) and oleoresin (3-13 per cent). Curcumin is the substance that is responsible for the biological activity of turmeric. Turmeric rhizome is 70 per cent carbohydrates,7 per cent proteins,4 per cent minerals and 4 per cent oil. It also has vitamins and alkaloids. Curcumin is used in cosmetics, preservatives, food industries and pharmaceuticals. The artificial colouring agents have been banned and so the use of curcumin is prompted. Curcumin protects the liver from toxic compounds as it acts like an anticoagulant by inhibiting collagen and by adrenaline induced platelet aggregation. Curcumin is also used to heal the wounds. It has antifungal and antiseptic effects.It has also antiviral effects and is found effective against the HIV. Various kinds of cancer including skin, colon and prostrate can be cured by the use of curcumin. It has been proved through the research that turmeric stabilizes and protects biomolecules in the body at the molecular level which is shown in its ant-oxidant, anti-mutagenic and anti-carcinogenic action.

Sugandham, Suvarna, Prabha, Rajendra Sonia, Narendra Haldi-1 *etc.* are the important varieties of turmeric. The planting material should be treated with Mancozeb (0.3 per cent) or Ridomil (0.25 per cent) before planting at least for 30 minutes. The planting material is stored in a cool dry place or in earthen pots plastered with mud. 2-6 ploughing is done before sowing the turmeric crop. 1-1.2mt wide beds are made with 15 cm height and convenient length. The ridges or furrows are used for sowing the turmeric. In the Northern region the turmeric is grown

in April and May. 60cm distance is kept between rows and 20 cm distance is kept between plants. Around 20qt. of disease free rhizomes are used for one hectare area. Turmeric crop requires heavy manures and fertilizers. Crop requires 25-30t/ha of well decomposed FYM or compost, 60 kg of Nitrogen, 30 kg of Phosphorous and 90kg of potash/ha. Full dose of phosphorous and potash are applied at the time of planting. The Nitrogen is applied in two split doses after 45 and 90 days after planting. After mulching the beds should be mulched with green leaves so that the soil moisture can be maintained and weeds can be controlled. In turmeric 3-4 weedings are required to get the desired production. The crop gets ready to harvest in the second fortnight of February. The complete drying of leaves is the sign of maturity. The rhizomes are carefully dugout from the soil and after digging the soil is cleaned from the rhizomes and is kept in cool and dry place for the post-harvest. The yield of turmeric varied from 200-250qt/ha.

The crop mostly is infested with the rhizome rot disease which is caused by the fungus namely *Pythium aphanidermatum*. The collar region of the pseudo stem becomes soft and water soaked resulting in the decay of the rhizomes. Seed treatment with 50gms of Mancozeb in 15 liters of water for 30 minutes prior to storage and at the time of sowing controls the disease. When the crop is infested in the field then the fields should be drenched with the mixture of Mancozeb and Carbendazim @3gms/litre of water to control this disease. Turmeric fingers are removed from the mother rhizomes and mother rhizomes are kept aside as seed material. Curing of rhizomes is done by boiling of fresh rhizomes in water for 45-60 and drying in the sun. Boiling is stopped when the froth comes out and white fumes appear giving out a typical odour. The boiling influences the aroma colour of turmeric and overcooking spoils the quality. The cooked fingers are then dried in the sun by spreading them in 5-7cm thick layers on the floor. The thin layer adversely influences the colour of turmeric. It requires at least 10-15 days to be completely dry. Artificial drying with the cross-flow hot air at a temperature of 60 degree C also gives the good quality and colour to the turmeric. The yield of the dry product varies from 10-30 per cent depending on the variety and location. If grown commercially, it can be a boon for the farmers of monkeys affected areas.

Chapter 25

Forest Fire Management in J&K

Union Territory of Jammu and Kashmir is one of the mega bio-diversity regions of India. J&K is abundant with the unique and diversified floral and faunal wealth. With 20,441 sq.kms of forest cover, J&K is one of the richest areas of bio-diversity in India. J&K ranks 12th in respect of area under forest cover in India. The forests of J&K are economically very rich and have great benefits. Jammu and Kashmir is a forest rich UT. About 19.95 per cent of the total geographical area of J&K is covered under forests. Around 59 per cent of the geographical area of J&K is under permanent snow cover, glaciers and cold desert and thus unable to support trees growth. The national level goal in forestry is to cover at least two thirds of the total area in the hilly regions under forests and trees cover. Jammu and Kashmir regions forests account for 47 per cent of the geographical area. Jammu and Kashmir adopted its own forest policy to maintain adequate forests in various regions of the region in 2011.Jammu and Kashmir is one of the most beautiful tourists' destinations in the world. Forests were generally vested with the government especially after 1947 but usufruct rights of local communities, ethnic groups and tribes continued to be part of management. Irrespective of ownership, forests were treated as open or community resources of local population in the limited sense of traditional rights. J&K is bestowed with different types of forests ranging from sub-tropical to moist temperate and alpines. Per capita forests and tree cover in J&K is about 0.15 ha which is almost double the national average i.e. 0.07 ha. Around 40 per cent of the forest area in J&K is either dense or very dense. 60 per cent of the total forest area falls in Jammu region and rest falls in Kashmir region. Forests play a significant role for the sustenance of livelihood of rural and urban people. Forests are also important for power development, tourism, agriculture, horticulture *etc.*

Damage done by the forest fire can be mild or severe. Fire can damage the forest trees, forest soil, and natural herbs, reduces the productive capacity of the forests, affects the wild animals, damages the scenic beauty of the forests, damages various other organisms in the forests *etc.* In order to manage the forest fire some factors are really needed to be studied. Weather needs to be properly studied which determines the fire environment. Temperature and winds are very important factors for the forest fire. With the climate change rainfall pattern is also changing. If intermitted summer rains through westerly depressions are not there then it can be a major cause for forest fires. Forest fire do not occurs unless there is inflammable material such as dry grasses, shrubs, fallen leaves and fallen woods are available in the forest. The more the inflammable material available in the forest, the more severe will be the forest fire. Topography is another factor for the forest fire in the hilly areas. Fire spreads very fastly when it travels towards a hill slope but its spread slowly when it travels towards downhill.

Generally people think forest fire as normal fire and expect a quick control over it but actually it is very cumbersome to control and extinguish it especially from hilly areas. The fire usually starts naturally or intentionally from a particular spot in the forest area and then due to wind, inflammable material and topography spreads fast in one or another direction. The speed, direction of spread, extent and shape of fire depends on wind, inflammable material and topography. If the fast wind is not blowing in the area at the time of start of the fire then it spreads in all directions slowly and thus assumes a circular shape. But if the wind is blowing in a particular direction then the fire spreads in that direction fastly and in other directions slowly. If inflammable material is available on a particular side then fire spreads in that particular direction fastly and slowly in other directions. The spread of fire in a hilly terrain is further affected by topographic factors, particularly the slope.

Fire is the most destructive element. It can destroy all forms of life, can cause soil erosion, kills all micro-organisms and destroys the ecosystem that have been built up over a long period of time. There is need to devise and execute the strategies to control the fire. Forest fire can be managed through different methods. Well equipped control rooms needs to be established with connectivity by proving wireless sets, cell phones, transportation facilities *etc.* to the fire fighters. Proper training on forest fire management should be imparted to the concerned staff and local people, first aid kits for emergencies and tools, creation of fire lines and maintaining them in the subsequent years. There is need of the quick action to suppress the fire and the delay can be avoided by quick detection and quick communication. There is need to deploy local fire watchers during the fire season and proper communication facilities should be provided to them. There should an arrangement of tools, food, water and light for the concerned people as the process of fire extinguishing can take some days. There are various conventional methods for fire extinguishing such as by water, by soil, by fire beating *etc.* Now-a-days helicopters are also being used in the state to extinguish the fire which is a good initiative. While extinguishing fire there is need to bear non- synthetic fire proof clothing to protect the body from heat, use head gear and goggles from radiant heat, use footwear preferably leather boots, carry sufficient water to guard against desiccation by heat, carry first aid kit for emergencies, to plan for escape route in case of danger, to maintain communication between fire fighters. Satellite data may be of quiet significance to evolve new innovative techniques in the field of forest fire management. Along with the satellite data many field-based attempts can also be made at various levels to collect forest fire statistics in the state. There is need of serious efforts to overcome the problem and tackle the problem of forest fire in a more effective way.

A forest fire causes irreparable damage to wildlife, biodiversity, water resources, forest based livelihoods and well being and changes heaven into hell within very less time. Fire control measures such as creation of fire lines, effective information and communication technologies (ICTs), defined responsibilities and introducing the incentives to the local communities living in the fringes of these forests for participation in the fire management must be developed. Training programmes with regard to fire fighting to government employees, local communities and interested rural youths/persons in order to create more awareness on causes and the control of forest fires should be arranged. Watch towers and other fire detecting systems should be developed in the Union Territory of J&K. Awareness campaigns on fire damage, prevention, detection and communication and suppression should be launched by involving schools, Joint Forest Management (JFM) committees, Non-Government and other voluntary organizations well before the onset of fire season every year. Wide publicity on forest fire management should be done through pamphlets, handouts, circulars, posters, TV, radio, video tapes *etc.* A fire management research and training institute for foresters and public should be established in J&K for providing knowledge and skills. The forest officials are required to rehearsal and practice the fire suppression exercise involving the local community. Fire experiment stations should be set up in representative regions of J&K to study various effects of fires in different ecosystems and provide extension services to the concerned areas. Adequate financial support on forest fire management should be provided to the concerned departments. Research and Development (R and D) institutions dealing with forest fire management should be strengthened. There is need of coordination of forest department with other agencies, whose support may be very important in forest fire management. Various seminars, training programs, conferences, study tours *etc.* should be organised on forest fire management. Awards for exemplary works in forest fire prevention, protection and suppression should also be distributed among the firefighters. Modern and effective tools and machineries for firefighting should be promoted.

Chapter 26

Women Empowerment through Self-Help Groups (SHGs) in J&K

Women constitute around half of population in the social system. Although women have made progress all the spheres of life but still there exists a lot of gender discrimination in the society. Poverty and gender discrimination threaten the well-being of women, deny their choice for education and training, restrict them for their participation in different programmes, curtail political and economic rights *etc.* In many cases women are given secondary importance in every section of society. Right from beginning of her life restrictions are imposed and she is prohibited to express herself properly. It has been observed that large percentage of population of women is unemployed, have poor socio-economic status, least decision making, lack education and awareness on different developmental aspects and so on. So, there is dire need to empower the women through different kinds of schemes or programmes launched by different governmental and non-government organizations. Some strenuous efforts are needed specially to empower the farm women.

Self-Help Group (SHG) is a small, economically homogenous and affinity based group of people who decide to save and contribute to a common fund to be lent to its members as per the group decisions. SHGs are a way to involve rural men and women actively in the developmental process by increasing the income, technical skills, mutual help, address constraints *etc.* The agenda of SHGs is 'Empowerment' and shift from dormant masses of the rural women to vibrant masses by bringing more income in their hands so that they can lead a better social life. Self Help Groups (SHGs) formation in J&K has immense scope. It especially in hilly areas of J&K can assist in the women's empowerment. The origin of Self-Help Groups is the brain child of Economics Prof. Mohammad Yunus of Chittagong University who founded Gramin Bank of Bangladesh in the year 1976.This was exclusively established for the poor section of the society. The Self-Help Group movement is a silent revolution to uplift the poor people across the globe. World Micro Credit Summit was held at Washington in 1997, converged the developed and developing countries of the world to overcome the serious problem of poverty by using micro credit as a tool to empower the poor section of the society. The major objectives of SHGs are: (1) To inculcate the habit of saving and banking among the rural women. (2) To develop credibility among the rural women and the bankers. (3) To develop group activity so that women can start the entrepreneurships. With the introduction of some new interventions micro-entrepreneurship can be developed among the rural women SHGs.

Sex ratio in Jammu and Kashmir is 889 i.e. there are only 889 females per 1000 males much below the national average of 940 females per 1000 males. The economy of Union Territory of Jammu and Kashmir has been facing an armed conflict from the last 30 years. Disturbance in this UT has deeply affected people's employment, their attitude, behaviour, habits, health *etc*. The women of Jammu and Kashmir have suffered a lot especially after 80s.Women can contribute significantly to the economy of the Union Territory. Women are the key agents for the society development. Government have launched a lot of state and centrally sponsored schemes in the J&K from time to time for the upliftment of women. About 3.8 lakh of women in J&K have involved in the SHG movement.

It has been observed that poor people have innate capabilities and make strenuous efforts to come out from poverty. Self-Help Groups (SHGs) especially for women help to start entrepreneurial activities and thus can help to eradicate the poverty. Formation of Self-Help Groups can be an effective tool for the women empowerment in the Union Territory. With the help of Self-Help Groups (SHGs) women can be trained for different kinds of skills under different programmes and schemes by the various organizations which further can help them to improve their socio-economic status. The concept of women's empowerment is the result of several important critiques, debates and discussions generated by the women's movement across the globe, especially by the third world countries.

Women and SHGs in many parts of the country have achieved success in bringing the women to the main stream of decision making. The SHGs in our country have become a source of inspiration for women's welfare. Formation of SHG is a viable alternative to achieve the objectives of rural development. SHG is also a viable organized set up to disburse micro- credit to the rural women and encouraging them together into entrepreneurial activities. To alleviate the poverty and to empower the women, the micro-finance, Self-Help Groups (SHGs) and credit management groups have also started in J&K. There is no doubt that the women of Jammu and Kashmir have suffered a lot during the last 30 years. SHGs formation can be one of the important options to assist these women to come out from economic distress. For this purpose, government has launched several state and centrally sponsored schemes/programmes for the upliftment of women. Some of

WELCOMES

these are as: (1) Women belonging to minority community *viz.*, Buddhists, Sikhs, Christians *etc.* are being financed under the national minorities' development and finance corporation.(2) Swarnjayanti Gram Swarozgar Yojana (SGSY) scheme was introduced in 1999 with the major objective of developing micro enterprise in rural areas, thereby developing the potentials of rural poor who belong to below poverty line, financial assistance both in the form of loan and subsidy is provided to the beneficiaries, both individuals as well as Self-Help Groups so that they can set up their own entrepreneurships. As per the reports, the Swarnjayanti Gram Swarozgar Yojana scheme is implemented in all the districts of Jammu and Kashmir. This scheme mainly emphasis on the formation of women's Self-Help Groups. (3) Reports signify that Jammu and Kashmir women's development corporation (JKWDC) has put forth various socio-economic schemes for the upliftment of women belonging to minorities, backward and other classes.(4) Various social welfare departments also implement various schemes for the women. Presently, there are around 150 social welfare centers which impart different kinds of training to the women so that they can start their entrepreneurships. Presently there are 19 Krishi Vigyan Kendras (KVKs) under SKUAST-J, SKUAST-K and CITH in the J&K which are imparting need based farmers and vocational trainings to the women and the trained women start the entrepreneurships in agriculture and allied sectors. These KVKs assist in the formation, management and strengthening of SHGs. Many women trained by the KVKs have started their micro-entrepreneurships in agriculture and allied sectors and generating a good income.

The size of an ideal Self-Help Group (SHG) should be of 10 to 20 members. An informal group should not have more than 20 members and the group members should have homogeneity in their socio-economic status. It has been observed that the members in a big group do not actively participate and the chances of conflict also increase. Only one person from a family should become a member of a Self-Help Group (SHG) and in this way more number of families can join Self-Help Groups. The group should be either of only men or only women. The group leaders viz. chairman, secretary and treasurer are decided by the group members. The group leaders should be benevolent in nature and they can be rotated as per the need. It has been observed that the women groups generally perform well and women are considered more credible by the banks for returning the group dues than their counterparts. The group members should follow rules and bye-laws of SHGs. The group should meet at weekly or fortnightly intervals and the participation of all the group members in all the group meetings makes easy to stabilize and develop credibility among the group members. Fine can be fixed for not attending the group meetings and after sometimes interloaning can be initiated in the group. Registers of the group should be kept upto date by the group by making the entries regularly. The funds should be managed and controlled by the group members themselves without any outside interference. Self-Help Groups are mainly based on savings by the group members and credits from the banks. Savings and credit services from local banks are the logical methods of extension of SHGs growth strategy to mitigate the increasing credit demand in the group. The major objective of Self-Help Group is to economically empower the group members. The members join the group on voluntarily basis. They save the money and then link their credit to a nearby bank to start an economic activity. Group members are guided by government or non-governmental organizations (NGOs). Women Self-Help Groups in Reasi distt. on milk processing, Parshad making, food processing, bracelets making *etc.* are assisted by the Krishi Vigyan Kendra, Reasi, local banks and NGOs and are working excellently. The successful SHGs can be a source of inspiration for the other people who want to earn a substantial income for the upliftment of their socio-economic status. The Self-Help Groups can be an effective technique for women empowerment in the society.

Chapter 27

Attracting Rural Youths of J&K in Agriculture

According to the UN youth is defined as a person in 15-24 years of age group whereas the census of India treats people in the age group of 15-29 years as youths. The participation of the rural youths for development in J&K can be more impressive if more attention is paid on them. According to 2011 census sex ratio in Jammu and Kashmir is 889 i.e. there are only 889 females per 1000 males much below the national average of 940 females per 1000 males. The economy of Union Territory of Jammu and Kashmir has been facing an armed conflict from the last 30 years. Disturbance due to militancy in J&K has deeply affected people's employment, their attitude, behaviour, habits, health *etc.* The rural youths of Jammu and Kashmir have suffered a lot especially after 80s and this process is continued. Rural youths of the J&K can contribute significantly to the economy of J&K. Rural youths are the key agents for the development of the J&K. Government have launched a number of state and centrally sponsored schemes in the state from time to time for the upliftment of rural youths. It has been observed that rural youths of the J&K have innate capabilities to change the destiny of this U.T. Rural youths are the future of the agricultural sector in J&K. A large number of rural youths of the Union Territory of J&K are migrating towards the urban areas in search of employment and to live a better life. They are migrating towards urban areas because there are limited resources, poor infrastructure and unemployment in the rural areas. In this way we are losing the present and future agriculturists in this U.T. Agriculture is an important tool which can stop the migration from rural areas towards the urban areas and can create the income and employment opportunities for the rural youths.

Youth is often the time when one starts to dream of the future, thinks of the path to take and boldly and aggressively set the life in motion. In many villages, to be a farmer is not a part of this dreamt future. What will then be the future of agriculture without the involvement of young farmers? If there is no farmer then there will not be food and life. It has been observed that the majority of rural youths are not attracted towards agriculture. We need to find out the initiatives being taken to encourage the youth to be in agriculture and need to give the recommendations for the youth to stay in this profession. India is said to be the youngest country in the world as it is having the highest number of youths in the world. The rural youths are often unemployed or work informally in unpaid or underpaid, low skilled and insecure. The lack of opportunities and decent jobs in the rural areas compel the youths to migrate to cities. For many decades, the rural youths in the state have been under-tapped and neglected by their communities, governments and international organizations and thus are unable to make full use of their energy and potentials in the agricultural sector. Sons and daughters of the farmers in J&K are often reluctant to go into farming due to various reasons. Many of those who continue to stay in the farming were often forced or did not have better options. This has resulted to ageing of the farming population. The youth is the future of the J&K and the rural youth is the future of agriculture and rural industry. How can we then attract the youth to stay or work in the rural areas? Different kinds of policies and programmes are needed to make by the government for their secure future and stay in agriculture and rural industry by their choice.

Agriculture provides the main source of income in J&K. Engaging rural youths in agriculture has been prominent agenda of the Government recently as there is growing concern across the nation that rural youths have become disenchanted with agriculture. It is predicted by the UN that by 2030, 6 out of every 10 people will live in the city and by 2050, this proportion will increase to 7 out of 10 people meaning that more young people than ever before are moving to cities and towns to find work, leaving few behind to work in the rural areas. It is easier to understand why the number of young farmers is in decline. So, how do we can reignite the love of rural youths for farming when the trend is to live in cities and towns is increasing?

It has been observed though various consultations that the rural youths in J&K are not attracted towards the agriculture. The rural youths think that agriculture is unglamorous, unprofitable, having less pride and dignity, low identity and back breaking. Farmers tell their children, "Do not be like me, just a lowly farmer." "If you don't study well, you will just end up here in farming". Many school children dream to be doctors, engineers or lawyers, but seldom like to be farmers. Farmers are always ranked low in the society and are defined as uneducated or uncivilized. Farmers are generally poor. They do not gain enough income from agriculture and are far from meeting their family, even personal, needs. Decreasing incomes due to increasing cost of agricultural inputs, such as seeds, fertilizers and pesticides, low prices of produce, lack of control over the markets and high risks in agriculture brought by unpredictable weather conditions and price fluctuations keep the youths away from agriculture.

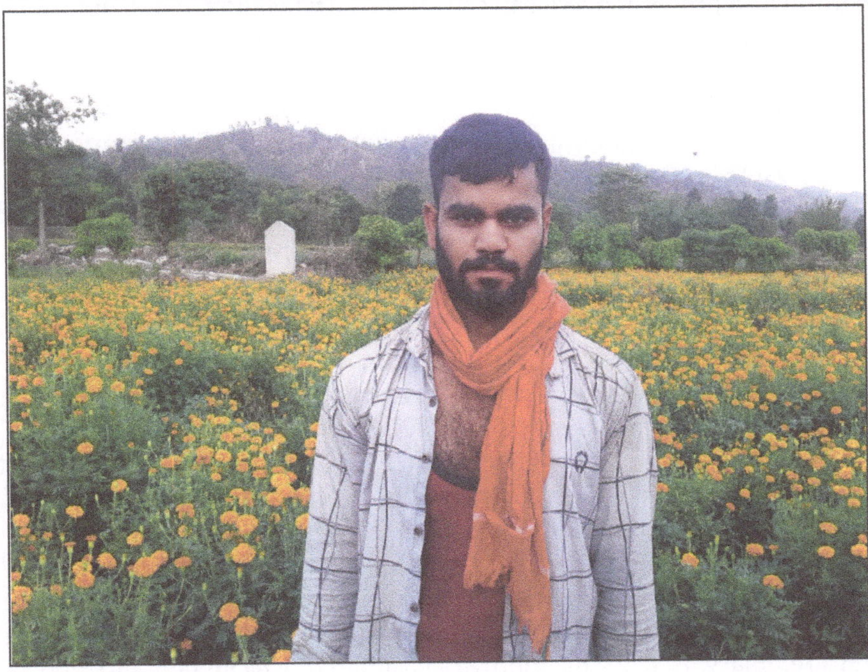

Ex-Prime Minister of India, Sh. Lal Bahadur Shastri gave the slogan "Jai Jawan, Jai Kisan" to stress that Jawans and Kisans are the two pillars of the nation. The future belongs to the nations with grains and not guns. For young people to take to agriculture, farming must be both intellectually satisfying and economically

rewarding. This will call for a technological and managerial upgradation of farm operations. We need to persuade the rural youths to stay in the villages and adopt agriculture as a profession. We can attract the rural youths in agriculture by the appropriate land use policies, technologies, market linkages, by enlarging the scope for the growth of agro-processing, agro industries, agribusiness and establish "Farm to Home" chain in production,processing and marketing. The Government of India has also launched a programme for the agricultural graduates to start agri-clinic and agri-business centres. Agri-clinics provide the services needed during the production phase of farming while the agri-business centre will cater the needs of the farm families during the post-harvest phase of agriculture. Thus, the farmers can be assisted in the entire crop cycle starting from sowing and extending up to value addition and marketing.

In order to ensure the food security in the country there is dire need to attract the rural youths in agriculture. They should be encouraged to take up agriculture and allied sectors enterprises for sustainable income and gainful employment. They should be trained for the capital intensive activities like processing; value addition and marketing.They should be guided to have links with different institutions and stakeholders for convergence of opportunities available under various schemes and programmes for the sustainable development. Krishi Vigyan Kendras (KVKs) established in the different districts of J&K are developing the skills on different entrepreneurial activities such as mushroom, poultry, goatry, vermicomposting, floriculture, apiary, seed production, dairy farming *etc.* Various demonstration units on agriculture and allied sectors established in the KVK are also helpful to train the rural youths. The agriculture and allied sectors model units are also established by the KVKs in the villages to attract the rural youths in agriculture. Skill development in agriculture will help the rural youths to improve their confidence, to pursue farming as a profession, generate additional income and employment opportunities.

Chapter 28

Uniqueness in Bhaderwah Rajmash

Jammu and Kashmir is endowed with diverse agro-climatic conditions. This hilly Union Territory is blessed with immense natural beauty. J&K is endowed with ample natural resources including soil, water, climatic condition, diversity, topography, rich natural flora *etc.* which are conducive for the cultivation of a wide range of pulses. Agriculture is the backbone of Jammu and Kashmir and this sector provides employment directly or indirectly to around 70 per cent of J&K population. Agriculture contributes about 65 per cent of the J&K revenue which signifies the over-dependence of this UT on agriculture. The average size of land holding of J&K is only 0.59 hectare against 1.33 hectares' land holding size on national basis. About 30 per cent of the area of J&K is under cultivation. The agro-diversity of J&K varies from sub-tropical in Jammu region, and temperate in Kashmir region. The average annual rainfall of these two regions is 1069mm, and 660mm respectively while the average temperatures are 24.5 and 13.3 Centigrade respectively. There is an urgent need to increase the income and employment in agriculture and allied sectors for the farmers in J&K and this is possible by the promotion of diversification in agriculture. J&K 58 per cent area under agriculture is rainfed and remaining 42 per cent is irrigated. Maize is the major crop of the J&K but there is an immense potential of pulses production in this UT.

Bhaderwah is a unique place on the earth. It is bestowed with treasures of nature. Bhaderwah valley is surrounded by snow-capped mountains, evergreen vegetations and tumultuous transparent rivers. Rajmash is grown mainly in Doda, Ramban and Kishtwar districts of Jammu division of J&K. Bhaderwah Rajmash is famous across the globe because of its delightful taste. It contains exceptional cooking qualities, easily digestible and attracting colour. Its grains are small in size and are unique in taste and colour. Being a popular dish it is prepared in almost all the occasions across the region. A 100 grams of boiled Rajmash beans contain 140 calories, 5.7 grams of protein, 5.9 grams of fat and nearly 17.8 grams of carbohydrate. It also contains anti-oxidants and rich in soluble fiber. Bhaderwah Rajmash is exceptional in taste and flavour. It has high demand in the market. Bhaderwah Rajmash is a local pole type trailing variety. It is an herbaceous annual plant grown for its edible dry beans. The main marketing channels are brokers and local traders. They facilitate the farmers to sale their farm produce.

Rajmash in Doda district is mainly grown in Bhaderwah, Sungli, Chinta, Sartangal, Nalthi, Kakol Mathal, Kansar, Lanchan, Thathri, Gandoh and Premnagar. It is also grown in Paddar, Chatroo, Warwah, Dakshin, Chingam and Dehrna area of Kishtwar district and Gool Gulabgarh area of Ramban district. Farmers collect their own seeds from better genotypes and keep it for the next generation. Mostly good quality seed from previous year crop is collected and preserved for the next season. To observe the desired tilth, the land is ploughed with MB plough (soil turning plough) for 3-4 times. Each ploughing is followed by planking to ensure fine tilth and moisture conservation. Its crop is sown in a well drained soil, as Rajmash and maize both are sensitive to water stagnation. Manures and fertilizers both play an important role in Rajmash cultivation. It is slow in biological nitrogen fixation because of poor nodulation. Farmers add organic manure in the soil to improve the general physical conditions and structure of the soil. About 10-15 tonnes/ha well rotten farm yard manure or compost is applied by the farmers before sowing and it gives good results. Additional doses of N, P and K are also applied by the farmers for its better yield.

Rajmash is a very popular crop in the area. Rajmash and rice is the most likely dish in the North India. Rajmash and rice of Peerah in Ramban district are very famous and going further the Rajmash of Chinta Valley in Doda district are said to be amongst the most popular. Traditionally it is grown as an intercrop with maize from since ages. This intercrop combination is an ideal combination to generate high economy in the hilly regions and is a sustainable soil enhancing practice. The local tall white maize variety gives ideal staking to its plant for its twining pole type variety. It is harvested in 2-3 pickings at 7- 10 days interval as the pods at lower end mature earlier and seeds are dispersed by over maturity in the field. Maize-Rajmash combination also helps to reduce soil erosion of the sloppy lands in the hills, which is a major problem of this area particularly in the solo maize growing fields. This is a traditional agricultural practice adopted since ages in this region. Bhaderwah Rajmash is a pole or trailing type variety of 3-4 m long.

It is sown in April-May. Seed Rate of 8 kg/ha of Rajmash is used when taken as intercrop with maize (30-35 Kg/ha)(2: 1).Row-to Row distance is 30 cm and plant to plant distance is 15-20 cm. Sowing depth is 5-8 cm. Before sowing, seed treatment is done with Bavistin or Thiram @ 2gm/Kg to protect the crop from various diseases at early seedling stage. The sowing is done when there is proper moisture in the soil. Rajmash is severely affected by the weeds in initial stages. First hoeing is done 30-45 days after sowing. Thinning is accompanied with hoeing and weeding after 25-30 DAS. Chemical weed control in Rajmash is also done by the

application of weedicides. Bhaderwah Rajmash varieties require support during the crop growth period. So when mixed with maize, it gets natural support from the maize plants. Harvesting of Bhaderwah Rajmash is done after 120-130 days after sowing when leaves and pods turn yellowish brown. The upper leaves remain green till final harvest. Anthracnose (seed and soil borne) disease in Rajmash is managed by the seed treatment with Bavistin (0.2 per cent) with 0.4 per cent formulations of T.viridie. The Rajmash harvesting is completed in August to September before the harvest of maize crop. The harvested crop is kept for sun drying for a week and thereafter thrashed. After thrashing, seeds are further cleaned and dried for storage. Storage is done in tin bins for 1-2 years. Its yield in mixed cropping is 6-8 qtls./ha.

Rajmash kidney beans help to maintain the metabolism of amino acids, gluconeogenesis, neurotransmitter synthesis, fatty acids, lipids, histamine synthesis and hemoglobin synthesis. It is very useful to maintain a healthy skin. There is great demand of Bhaderwah Rajmash across the world. So, there is need to enhance the area, production and the productivity of this important pulse in the district as it is an excellent source of earning for the farmers of the area. Quality research and irrigation facilities are also needed for the Rajmash growers in the district. There is also need to evaluate the possibility of Organic Rajmash in the district for improving the income of the farmers. Farmers Producer Organic Producers (FPOs) of Rajmash growers can be made. Endevours are required for mechanizing the production and introducing new marketing interventions for the Rajmash growers of the district. Mechanization production and introduction of marketing interventions are also required by the farmers. Rajmash growers of the district require special financial support and expertise to protect the future of the Rajmash cultivation in the district. Special policies are needed to strengthen the Rajmash cultivation in the district.

Chapter 29

Possibility of Strawberry Cultivation in J&K

Reasi district of J&K is primarily known for the cultivation of field crops such as maize, wheat, mustard, chickpea, turmeric, garlic, onion *etc.* This is a hilly district and about 94 per cent area of the district is rainfed. The soil of the district ranges from sandy loam to loam. The district receives average annual rainfall of 1100 mm. All the rainfall drains and forms uncontrolled flow which ultimately falls into the Chenab river which is considered as a big water body in the district. This district has varied climate ranging from temperate to sub-tropical. Rainfall is heavy and well distributed from June-September. Rainfall is moderate in other months. Sometimes dry spell also prevails in the district. Mountainous terrain and high peaks in the district are the main factors for the climate variation in the district. Summers are generally warm and winters cold with snowfall in higher altitudes. There are very less water harvesting units in the district. This district can be considered as the gold-mine of high-value horticultural crops. The size of land holding is very small and crops productivity is low. The income of the farmers is low. So, there is need to explore new options for the farming community so that their income can be enhanced and livelihood can be sustained. It was observed by the KVK, Reasi scientists that strawberry can be a good venture for the farmers to increase their income and employment. Strawberry has immense scope in the district as there are many religious spots in the district and tourists visit throughout year in the district. There are also many police and army camps in the district who can also purchase the farm produce. It was observed that strawberry can transform the socio-economic status of the farmers of hilly areas.

Strawberries are the popular soft fruits liked by everyone. Over the years the consumption of strawberries as fresh fruit has increased. Strawberries can be grown in the plains as well as hilly terrains at an elevation of 3000 m. Strawberries may offer quickest returns to the farmers in the shortest possible period. Strawberry is rich in proteins, calcium, phosphorus, potassium and vitamins such as A, B1, B2 and C. It is considered as the natural source of antioxidants. It is also anticarcinogenic and anti-diabetic. It can also be used in making jam, jelly ice cream, soft drinks *etc*. Strawberry is a low growing plant and is planted at a spacing of 60x30 cm accommodating 50-60 thousands plants/ha. One ha field can yield about 150 qt/ha of strawberry fruit. From one ha of area of strawberry, farmers can generate about Rs.8 lakh within six months. Strawberry crop is care intensive and requires irrigation at regular intervals. Honeybees are the main insect pollinators. It is a short stature and short duration crop. It can be sown in the inter spaces of orchards. By doing so farmers can get the additional income. It is in tune to doubling farmer income though horticulture. Most of the cultivars of strawberry produce hermaphrodite flowers and are self fertile. However, some also produce male or staminate, imperfect and pistilate flower which requires cross-pollination. If proper care in planting and application of irrigation water is taken, then it does not interfere in flowering process in the other orchard crops. The other *Rabi* crops namely wheat, coriander, mustard *etc*. which even in winter requires irrigation whereas orchard crops such as mandarin has no water requirement and they interfere with the flowering process which usually appears during February-March.

Sh. Jatinder Singh and S. Gurdev Singh (both are retired Govt. employees) progressive farmers from Bhabber and Bharakh village of Reasi district have given a new ray of hope for the other farmers of Reasi distt. by growing the strawberry in a scientific way. The villages of both these farmers are endowed with the climatic conditions suitable for strawberry crop. They were guided by the KVK, Reasi and SKUAST-J scientists to grow the strawberry commercially. They were guided by the scientists to cultivate the strawberry under polythene mulching sheet as the yield under polythene mulching sheet can be obtained higher than open conditions. Mulching in strawberry minimizes the freezing injury, suppresses weed growth and more importantly reduces the chances of softening of fruits. The Chandler variety of strawberry was suggested as it gives higher yield and is suitable in the area. Organic package of practices were also suggested to the farmers to grow the organic strawberry. They were also educated on the vermicomposting technique and organic formulations. Organic inputs for strawberry cultivation were provided by the KVK, Reasi scientists. Literature on organic farming was also provided to them. They were guided to grow the strawberry on raised bed by using the polythene sheets.

Now they are selling their strawberry in the local markets even in the lockdown period. Reasi Distt. administration is facilitating to sale their farm produce at a good price. They expressed that they could had earned more provided planting material, packaging material and small scale industry had been made available locally. They further opined that at present they are lacking the packaging material and if attractive packaging material is provided to them they can earn more money.

Many other farmers are showing inclination towards strawberry cultivation. As strawberry is a new crop in the area, farmers have shown keen interest in its cultivation. This crop can help the farmers to double their farm income. There is need to develop strawberry hubs in the district so as to take this crop on a larger scale. Strawberry cultivation can help in the holistic development of the farmers of the district.

References

Ahamad, B.H., Bodha, R.H. and Wani, A.H. (2010). Nutritional and medicinal importance of mushrooms, *Journal of Medicinal Plants Research,* 4(24): 2598-2604.

Ahamad, Shahid and Lal, B. (2019). Organic farming in J&K State, India. Current Investigations in Agriculture and Current Res. Lupine Publishers. Vol. 7, Issue 3, pp. 958-960.

Ahamad, Shahid and Sharma, J. (2019). Transformation of Indian Agriculture through Innovative Technologies. Daya Publishing House, New Delhi.

Ahamad, S. and Narain, U. (2007). Ecofriendly Management of Plant Diseases Published by Daya Publishing House,New Delhi, pp. 412.

Ahamad, S. (2013). Hill Agriculture. Published by Astral Publishing House, New Delhi, p. 550.

Ambastha and Singh, K.N. (1978). Intera-personal communication pattern of extension personnel, *I.J.E.E.,* 14(1 and 4): 8-15.

Bazaz, N.H. and Haq, I. (2013). Crop Diversification in Jammu and Kashmir: Pace, Pattern and Determinants. *IOSR Journal of Humanities and Social Science (IOSR-JHSS).* Volume 11, Issue 5 (May. - Jun. 2013), pp. 01-07

Centre launches organic farming scheme in J&K, Times of India, Oct 19, 2014.

Chand, R. (1996). Diversification through High Value Crops in Western Himalayan Region: Evidence from Himachal Pradesh. *Indian Journal of Agricultural Economics,* Vol. 41(4): 432-441.

Chaube, H.S. and Ramji Singh (2001). Introductory Plant Pathology. IBD Co., Lucknow, 360 pp.

China, R. (1986). Nigeria: Training and demonstration: A system to meet a need; PRDC Bulletin: University of Reading September.

Darzi, M.I. (2016). Horticulture sector towards economic development of Jammu and Kashmir. *International Journal of Multidisciplinary Research and Development.* Volume 3; Issue 4; April; Page No. 238-240.

Das, S.K. and Itnal, C.J. (1994). Capability based land use system: role in diversifying dryland agriculture in dry area. *Bull Indian Soc. Soil Sci.* 16: 92-100.

Dhondyal, S.P. (1997). Farm Management: Economic Approach, pp. 213-216.

Ganga, G. and Chetty, J.S. (1971). An introduction to sericulture. Oxford and IBH Publishing Co. Pvt. Ltd., New Delhi.

Gupta, R.D. (2015). Agricultural Diversification. *Daily Excelsior.*, 2.9.2015.

Jana, B.L. (2003). Diversification in agriculture, Mohisarw Book Series, Srinagar.

Jha, D. (1996). Rapporteur's Report on Diversification of Agriculture and Food Security in the Context of New Economic Policy, *Indian Journal of Agricultural Economics* Vol. 51, 829-832.

Joshi, P.K., Gulati, A., Birthal, P.S. and Tewari, L. (2004). Agricultural Diversification in South Asia: Patterns, Determinants and Policy Implications. *Economic and Political Weekly*, Vol. 39, 12-18.

Kamili, A.S. and Massodi, M.A. (2000). *Principles of Temperate Sericulture*. Kalyani Publishers.

Khajuria, A. and Jain, D. (2016). Organic Farming Potential as Sustainable Agriculture: A J&K Perspective Volume 2, Issue 4, pp. 257-260.

Lal, B. and Sharma, P. (2019). Potential of commercial poultry farming in J&K. *Early Times*. 11.15.2019.

Lal, B and Sharma, P. (2018). Challenges and Opportunities in Organic Farming. *Early Times*. 9.12.2018.

Lal, B. and Tandon, Vikas (2015). Horticulture sector in J&K. *Daily Excelsior*. 8.11.2015.

Lal, B. and Tandon, Vikas (2019). Blooming floriculture in J&K. *Daily Excelsior*. 7.05.2019.

Lal, B. (2016). Potential of diversified agriculture in J&K. *State Times.*

Masoodi, M.A. (2006). Agriculture in J&K: A Perspective. Agrotech Publishing Academy, Udaipur.

Pingali, P.L. and Rosegrant, M.W. (1995). Agricultural commercialization and diversification: Processes and policies. *Food Policy*, Vol. 20, 171-186.

Sartaj, A. Wani, Chand, Subhash, Najar, G.R. and Teli, M.A. (Research Journal on Organic Farming as a climate change adaption).

Sawant, S.D. and Achuthan, C.V. (1995). Agricultural Growth across Crops and Regions: Emerging Trends and Patterns. *Economic and Political Weekly*, 30: 17-25.

Singh, K. (2019). Making sericulture a self-sustaining enterprise in J&K. *Daily Excelsior*. 10.10.2019.

Sud, A.K., Paul, Y.S. and Thakur, B.R. (1999). Corm rot of saffron and its management, *Journal of Mycology and Plant Pathology*, 29: 380-282.

Vyas, V.S. (1996). Diversification in Agriculture: Concepts, Rationale and Approaches. *Indian Journal of Agricultural Economics*, Vol. 51, 430-441.

www.dailyexcelsior.com Article on potential of organic farming in hilly areas.

www.thealternative.in Organic Farm Fresh: Those who quit agriculture are coming back through organic farming.

Zargar, B.A. (2013). Saffron production in Kashmir. *Greater Kashmir*. 13.10.2013.

Index

9 789354 614330

Printed at Repro India Ltd.